European Borderlands

The expectations of European planners for the gradual disappearance of national borders, and the corresponding prognoses of social scientists, have turned out to be over-optimistic. Borders have not disappeared – not even in a unified and predominantly peaceful Europe – but rather they have changed, become more varied and, in a certain sense, mobile, taking on an important role in the everyday lives of more people than ever before. Furthermore, it is now widely accepted that borders do not just hinder communication and the formation of relationships, but also channel and prefigure them in a positive way. Presenting a number of studies of everyday life in European borderlands, this book addresses the multifarious and complex ways in which borders function as both barriers and bridges. Focusing on 'established' Western European borderlands – with the exception of three contrasting cases – the book attempts a turn from conflict to harmony in the study of borderlands and thus examines the more mundane manifestations of border life and the complex, often unconscious motives of everyday cross-border practices. The collection of chapters demonstrates that even in the case of 'open' political borders, the border remains an enduring factor that is not adequately described as either a problematic barrier or a desirable bridge. The studies look at bordering processes, not only approaching them from different disciplinary angles – sociology, anthropology, geography, history, political science and literary studies – but also choosing different scales and making comparisons that range from different borders of one country to the reactions and attitudes of different individuals in a single borderland village.

Elisabeth Boesen is Senior Researcher at the University of Luxembourg's Institute of History, Luxembourg.

Gregor Schnuer is Research Associate at the University of Luxembourg's Institute of History, Luxembourg.

Border Regions Series

Series Editor: Doris Wastl-Walter, University of Bern, Switzerland

In recent years, borders have taken on an immense significance. Throughout the world they have shifted, been constructed and dismantled, and become physical barriers between socio-political ideologies. They may separate societies with very different cultures, histories, national identities or economic power, or divide people of the same ethnic or cultural identity.

As manifestations of some of the world's key political, economic, societal and cultural issues, borders and border regions have received much academic attention over the past decade. This valuable series publishes high quality research monographs and edited comparative volumes that deal with all aspects of border regions, both empirically and theoretically. It will appeal to scholars interested in border regions and geopolitical issues across the whole range of social sciences.

For a full list of titles in this series, please visit www.routledge.com/geography/series/ASHSER-1224

Informal Trade, Gender and the Border Experience
From Political Borders to Social Boundaries
Olga Sasunkevich

Borderscaping
Imaginations and Practices of Border Making
Edited by Chiara Brambilla, Jussi Laine, James W. Scott and Gianluca Bocchi

Russian Borderlands in Change May 2016
North Caucasian Youth and the Politics of Bordering and Citizenship
Tiina Sotkasiira

Transnational Frontiers of Asia and Latin America since 1800
Edited by Jaime Moreno Tejada and Bradley Tatar

The Politics of Good Neighbourhood
State, Civil Society and the Enhancement of Cultural Capital in East Central Europe
Béla Filep

European Borderlands
Living with Barriers and Bridges
Edited by Elisabeth Boesen and Gregor Schnuer

European Borderlands

Living with Barriers and Bridges

Edited by Elisabeth Boesen and Gregor Schnuer

Routledge
Taylor & Francis Group
LONDON AND NEW YORK

First published 2017 by Routledge

2 Park Square, Milton Park, Abingdon, Oxfordshire OX14 4RN
711 Third Avenue, New York, NY 10017

Routledge is an imprint of the Taylor & Francis Group, an informa business

First issued in paperback 2018

British Library Cataloguing in Publication Data
A catalogue record for this book is available from the British Library

Library of Congress Cataloging in Publication Data
Names: Boesen, Elisabeth, editor. | Schnuer, Gregor, editor.
Title: European borderlands : living with barriers and bridges / edited by Elisabeth Boesen and Gregor Schnuer.
Description: New York, NY : Routledge, 2016. | Includes bibliographical references and index.
Identifiers: LCCN 2016022520| ISBN 9781472477217 (hardback) | ISBN 9781315580852 (e-book)
Subjects: LCSH: Borderlands--Social aspects--Europe, Western. | Boundaries--Social aspects--Europe, Western. | Regionalism--Europe, Western. | Group identity--Europe, Western.
Classification: LCC JN34.5 .E8745 2016 | DDC 320.1/2094--dc23
LC record available at https://lccn.loc.gov/2016022520

ISBN: 978-1-4724-7721-7 (hbk)
ISBN: 978-1-138-62418-4 (hbk)

Typeset in Times New Roman
by Saxon Graphics Ltd, Derby

Contents

Illustrations

Figures

Maps

Tables

Contributors

Elisabeth Boesen is a senior researcher at the History Institute, University of Luxembourg, where she is currently directing a research project on 'Cross border residence: Identity experience and integration processes in the Greater Region'. Prior to this she worked as a researcher at the Centre for Modern Oriental Studies (Berlin) and at Bayreuth University. She holds a PhD in Cultural Anthropology from Bayreuth University. Her work centres on issues of spatial mobility and rural change in the West African Sahel and the Greater Region SaarLorLux.

Garance Clement is a PhD candidate at the University of Paris Est, at the Lab'Urba research unit. She researches middle class cross border residential mobility from an urban sociology perspective, conducting a case study in the agglomeration of Lille and the surrounding Belgian cities. She holds a MA in Political Sciences from the Political Sciences Institute of Lille (2011) and a MA in Urbanism with a focus on housing from the Urbanism Institute of Paris (2012).

Tamás Hardi is a senior research fellow in political and urban geography at the Institute for Regional Studies in the Hungarian Academy of Sciences. His research interests are the cross-border movements and region formation along the Hungarian-Slovak and Austrian borderlands, and the examination of the spatial perception and identities among the citizens of Central-European countries. He holds an MA in Geography and Adult Education (1994) and a PhD in Economics – Regional Studies (2002). He is a lecturer in Political Geography, Geopolitics and Central and South-Eastern European Studies at the Széchenyi István University.

Jarosław Jańczak is an assistant professor in the area of European integration at Adam Mickiewicz University, Poznań, Poland and the European University Viadrina, Frankfurt (Oder), Germany. He is currently working on cross-border governance in European border twin towns. He holds an MA in Political Science from Adam Mickiewicz University (2000) and a PhD in Political Science, with a focus on the enlargements of the European Union (2004).

Martin Klatt is Associate Professor of Contemporary History at the Centre of Border Region Studies, University of Southern Denmark. His research activities concentrate on national minorities in border regions, the Danish-German border region's history, regional cross-border cooperation, cross-border regions in history and today, and cross-border regional governance. He was PI for the SDU work package of the EuroCoreCode project "Unfamiliarity as signs of European times: Scrutinizing historical representations of otherness and contemporary daily practices in border regions" (2010–2014), co-funded by the European Science Foundation.

Nils Müller is a researcher and writing coach at Bielefeld University of Applied Sciences. His main research interests are everyday routines, theory of science and analytical sociology. In his dissertation research he analysed how routines of everyday mobility change in European border regions and how they are connected to processes of cross-border spatial integration. He holds diplomas in European Business Administration (2007) and Sociology (2008) from the University of Bamberg and a PhD in Sociology from the University of Oldenburg (2013).

Birte Nienaber is an associate professor in Political Geography at the University of Luxembourg, coordinator of the European Migration Network national contact point Luxembourg, steering group member of the UniGR – Center for Border Studies, and coordinator of the ESPON NCP Luxembourg. She is interested in the fields of migration, border studies and European regional development. She holds a Higher Diploma (2002) and a PhD (2005) in Geography from the University of Münster, and a Certificate in Higher Education (2010). She habilitated at Saarland University (2012).

Lesley Penné is an academic assistant in German at the Department of Language and Literature at the Vrije Universiteit Brussels. She is currently a doctoral candidate in German literature researching space and border concepts and Vergangenheitsbewältigung in contemporary German-language Belgian novels. She holds a Masters in French and Spanish Linguistic and Literary Studies (2008). She also completed her teacher training in 2009 and obtained a Master's in German Linguistic and Literary Studies in 2010 at the Vrije Universiteit Brussels.

Isabelle Pigeron-Piroth is a research associate at the University of Luxembourg. Her main research fields are employment and mobility (labour mobility as well as residential mobility) in cross-border contexts. Her research looks at cross-border labour markets and their impacts on cross-border development in the Greater Region SaarLorLux. She holds a Master's in Economics from the University of Nancy (1995).

Gregor Schnuer is an independent academic working on the research project 'Cross border residence. Identity experience and integration processes in the Greater Region.' He has previously worked as a research associate at the University of Luxembourg, course convener at the University of Edinburgh, and has taught at Napier University Edinburgh. He holds an MA in History and Sociology from the University of Edinburgh (2006) and a PhD in Sociology, with a focus on social theory, from the same institution (2010).

Keiu Telve is a PhD student in the Ethnology Department at the University of Tartu, and she holds a junior scholar position at the University of Eastern Finland. Her research interests include migration, cross-border mobility, gender and family issues. She is also one of the founders of the Center for Applied Anthropology of Estonia. She holds a Masters degree in Ethnology (2015) and a Bachelors degree in Literature and Cultural Research from the University of Tartu (2009).

Eva Zimmermann is a cotutelle PhD student at Freie Universität Berlin and Université de Lorraine in the subject area of history, with a focus on nineteenth century transnational cultural history. She holds a scholarship from the Land Berlin and is currently working on a thesis on "Baden-Baden – The Summer Capital of Europe: The Formation and Development of a Transcultural Space (1830–1880)". She holds an MA in History, Romance Philology and General and Comparative Literature form the Freie Universität Berlin (2011).

Abbreviations

ADULM	Agence de Développement et d'Urbanisme de Lille Métropole
BMW	Bondues, Marcq and Wasquehal
CBRM	Cross-Border Residential Mobility
CCSS	Centre Commun de la Sécurité Sociale, Luxembourg
CF de l'Est	Compagnie des chemins de fer de l'Est
DKK	Danish Krone
ESPON	European Spatial Planning Observation Network
EU	European Union
EU27	European Union including twenty-seven member states (Austria, Belgium, Bulgaria, Cyprus, Czech Republic, Denmark, Estonia, Finland, France, Germany, Greece, Hungary, Ireland, Italy, Latvia, Lithuania, Luxembourg, Malta, the Netherlands, Poland, Portugal, Romania, Slovak Republic, Slovenia, Spain, Sweden and the United Kingdom)
EuroCoreCode	European Comparisons in Regional Cohesion, Dynamics and Expressions (European Science Foundation research call)
GDP	Gross Domestic Product
IGSS	Inspection Générale de la Sécurité Sociale, Luxembourg
KERN	Kreis Rendsburg-Eckernförde and Stadt Kiel and Stadt Neumünster
NCP	National Contact Point
NPD	Nationaldemokratische Partei Deutschland (National Democratic Party Germany)
NUTS3	Nomenclature of Units for Territorial Statistics
OIE	Observatoire Interrégional de l'Emploi
PI	Principle Investigator
SDU	University of Southern Denmark
SODEVAM	Société de développement et d'aménagement de la Moselle
STATEC	Institut national de la Statistique et des Etudes économiques du Grand-Duché de Luxembourg
TV	Television
UniGR	University of the Greater Region
US	United States of America

Introduction

Living in European borderlands

Elisabeth Boesen

The 'Greater Region SaarLorLux', a loosely defined geographic area comprising the German federal state of Saarland, Lorraine in France and the Grand Duchy of Luxembourg, as well as parts of Belgian Wallonia and the German federal state of Rhineland-Palatinate[1] (and thus stretching across six national borders), is one of those border regions that seem to confirm European planners' vision of a Europe without borders, as well as postwar social scientists' analyses predicting the gradual disappearance of national borders. Traffic in the region is unimpeded; national borders no longer act as a barrier in the choice of workplace or even place of residence (Wille, 2012; Boesen and Schnuer, 2015).

In scholarship, however, a more complex understanding of borders and their functioning has evolved in recent decades, relativizing the optimistic expectation of the gradual disappearance of borders. Borders do not disappear – not even in a unified and predominantly peaceful Europe – but change, reappear, become varied, and in a certain sense become mobile, and thus they are a part of everyday life for more people than ever before (Newman, 2011, Balibar, 2003). Furthermore, Fredrik Barth's insight formulated with respect to ethnic groups has become widely accepted: borders do not just hinder communication and the formation of relationships, they also channel and prefigure them in a positive way (Barth, 1969; Newman and Paasi, 1998; O'Dowd, 2002). This more differentiated approach, which owes much to the integration of concepts and methods from social sciences (sociology, social anthropology, etc.), also helps researchers to investigate unexpected developments and to produce related hypotheses. One such unexpected phenomenon is the often-observed relative stagnation in the cross-border mobility of people after the opening of borders, especially after the Schengen agreements, as for example on the Dutch-German border.[2]

No such stagnation can be seen in the Greater Region SaarLorLux. Here, the opposite is the case: over the past decades, mobility has increased to an unexpected extent. Everyday life moves across borders for a very large number of inhabitants in the region, spurred by the enormous amount of cross-border commuters to Luxembourg and the existing disparities in excise duties and purchasing power, to name only a few factors. The use of open borders and the related perception of new opportunities reveal a variance that draws attention to the differences between the lived-in worlds of inhabitants of border regions. The editors of this volume intend to bring together studies of such everyday practices and experiences in a range of regions where national borders represent an important aspect of societal development and processes of cultural identity formation.[3]

The selection of cases is relatively small-scale and – with the exception of three contrasting cases – stays within the borders of the 'established Europe of the West' (Häkli and Kaplan, 2002, p.9). This restriction makes sense for heuristic reasons, since the absence of large historical-structural differences in the presented cases – such as violent political conflicts – leaves undisturbed the focus on the less-spectacular manifestations and the complex, often unconscious motives of everyday practices.

European borderlands

In the past two decades borders have become an object of research for scholars in a variety of disciplines, and they increasingly constitute a distinct field of research. The development of border studies has been accompanied by a fundamental conceptual reorientation: borders are no longer exclusively studied as territorial markings of the ambit of state sovereignty or other rule, but as an object and a result of continuing social processes as well as the cause of specific social processes in their vicinity (van Houtum and van Naerssen, 2002). This new, processual conception understands the actual object of border studies as the practices related to borders, summarised in the concept of 'bordering', with the variants 'de-bordering' and 're-bordering' (van Houtum, Kramsch and Zierhofer, 2005) – allowing significant differences regarding the constructivist understanding of these processes.[4]

All this also brings the spaces along borders into view. The previously more or less neglected margins have given rise to the new research field of 'borderlands', in which the everyday processes of de-bordering and re-bordering can be observed in unique manifestation and density, making them a kind of laboratory for scholars and planners. A large number of current border studies (including the contributions in this volume) deal with case studies on developments in individual borderlands. Oscar Martinez has suggested a typology of these on the basis of the prevailing modes of interaction. He differentiates between 'alienated', 'co-existent', 'interdependent' and 'integrated borderlands' (Martinez, 1994). Border studies are predominantly concerned with representatives of the third and the fourth types. Borderlands attract interest if they contain cross-border interactions, or if socio-economic and, above all, cultural differences become acute. The border is thus seen as exerting an effect which has an inherent potential (Minghi, 1991, p.15). Balibar expresses this when referring to border regions as centres where we are, in a particularly 'central' way, confronted with the question: 'Is there then "a European people", even an emergent one?' (Balibar, 2003, p.2). Lauth-Bacas and Kavanagh express the issue more clearly by talking – albeit using quotation marks – of a 'border culture' and a 'border identity' that 'takes into account the specific locality and proximity to the Other' (2013, p.10). However, this border effect does not necessarily ensue. Recent studies also show, as mentioned above, that the other can be met with indifference, even and especially if he (or it) has become ever more present due to facilitated cross-border exchange (van der Velde and van Houtum, 2004). Likewise the immediate political relevance of borders is viewed with scepticism.[5] This means that it is not clear from the outset what makes spaces on both sides of a border into borderlands in an emphatic sense.

The same applies to the derived category of 'borderlanders'. Martinez also provides a typology for those. He distinguishes 'national' from 'transnational' borderlanders, using the US-Mexican border to illustrate that these basic types are each composed of a multiplicity of variants. Similar typologies can be found in use by other authors. For example, Klatt differentiates between 'national borderlanders', 'border surfers', 'regionauts' and 'transnational borderlanders' (Klatt, in this volume). In this case, the comparative dimensions are even more complex than for borderlands, since they combine an actor-oriented perspective with structural aspects, especially the distribution of ethnic and linguistic identities across the borderland (Wilson and Donnan, 1998).

The majority of recent publications on borderlands can also be characterized through their dedicated geographical framings and the comparative approaches inherent to them.[6] As mentioned, this volume focuses on the 'established Europe of the West'. The majority of contributions deal with borderlands in the space that constituted the core of European integration: Benelux, France, Germany and – in a more limited sense – Denmark. They are complemented and contrasted by three borderland studies in the Eastern regions of the European Union.

As a result of the dramatic developments in other parts of Europe and the world, Western Europe has recently moved out of the focal point of borderland research. This is regrettable for several reasons – which in part echo Julian Minghi's call for a turn from conflict to harmony in the study of borderlands (Minghi, 1991). The regional integration processes have lasted longer here, and developments are therefore less marked by recent, more or less dramatic changes and conflicts, presenting themselves as relatively consolidated both in their successes and their difficulties. Seen on a global scale, these borderlands are also rather similar, despite important historical and structural differences. This relative similarity can be traced to long-term political and social developments as well as the economic and political efforts at integration that have taken place since the Second World War. We argue that these conditions allow for the production of particularly differentiated analyses. Individual case studies, each consisting of close-ups while also addressing very different spatial and social levels, complement each other and combine to form a comprehensive comparative analysis.

Furthermore, one can argue that western European borderlands, and especially those that are presented here, merit intensive study as they are 'overdetermined' borders. They are overdetermined not by virtue of representing civilizational differences or 'race relations' (Balibar, 2003, p.5), but rather in the sense that they epitomize the uniting quality of borders, their capacity for stimulating mobility and exchange. In particular, the Greater Region, with the highly symbolic locality of Schengen in its midst, would thus be a counterpart of the Balkan region, for example, where new borders have been established in recent years and border regimes are currently changing in dramatic ways. Contemporary research on borders and the ensuing theories are grounded in these complementary and somewhat exalted notions of the border.

The initial enthusiasm at the prospect of a world without borders has given way to disillusion. However, borders have not lost any of their fascination. On the contrary, in recent years borders have become a theoretically stimulating topic. This is proven by the continuous conceptual and methodological developments that are striving for new epistemologies and new ontologies (Parker, Vaughan-Williams et al., 2009; Parker and Vaughan-Williams, 2013) or are promising them as being encapsulated in the notion of 'borderscape' (Brambilla et al., 2015). However, critical voices can also be heard. Considerable scepticism about the idea of a general 'border theory' is expressed by Paasi: 'The challenge for border studies lies in the perpetual theorization of borders and boundary producing practices rather than in a solid border theory of some kind' (Paasi, 2011, pp.16–17). Accordingly he evaluates current attempts at 're-theorising' border concepts as an 'unending, self-feeding activity' (Paasi, 2014, p.364) and criticizes them as being influenced by normative views, which conceive borders fundamentally as regressive elements and hindrances to progressive politics (Paasi, 2014, p.368; Paasi, 2011, pp.19–22).

While conforming with the idea that it is important to study national borders as processual and multiple phenomena (Sohn, 2015), we also agree with the critical voices that highlight the context-boundedness of borders and the need for empirical research. This means that in our opinion, the right question to ask is not whether a differently founded border epistemology can be defined and new descriptive border ontologies be constructed (Parker and Vaughan-Williams, 2009), but rather how existing border ontologies – or border narratives, as Newman and Paasi (1998) put it – are to be grasped. In contrast to Newman and Paasi, we would contend that such narratives that are new in the sense of representing alternatives to dominant Western ideas of territorial boundedness are not only to be found within different cultural traditions (Rumford, 2012, p.889). The established Europe of the West, for example, offers them as well, and assembling variously scaled borderland 'close-ups' is a promising way to approach them.

Structure of the volume

The volume has a clear emphasis on the part of Europe that represents the historical core of the European integration process by bringing together seven articles on the French-German, French-Belgian, German-Dutch and German-Swiss border regions and on the Greater Region SaarLorLux. These studies are complemented by four papers on eastern and northern European borderlands. All of the articles are concerned with borders that, having been disputed historically and subjected to more or less dramatic shifts and changes, can be described today as uncontested and 'cold'. We would argue that although the borders in question have 'lost much of their drama' (Löfgren, 1999) and also much of their desire-producing capacity, they nonetheless involve interesting and diverse everyday processes that shape their respective borderlands. The following studies look at such processes not only by approaching them from different disciplinary angles – sociology, anthropology, geography, history, political sciences and literary studies – but also by choosing different scales and making comparisons that range from the different borders of one country to the reactions and attitudes of different individuals in a single borderland village.

The volume is organized in three parts. The first part is on everyday border crossings and the impact of border politics on these practices. The three authors of the individual chapters are all concerned with the question of why cross-border interaction takes place or, conversely, does not take place. They present different approaches to studying the diverse, and to a certain degree unexpected, developments of these interactions. The second part unites four contributions that deal with communities, relationships and identities that are shaped by shifting borders and cross-border practices. The last part of the volume is dedicated to one cross-border practice that has particularly far-reaching consequences for everyday life and thus deserves special treatment – namely, cross-border residential migration or, as one might say, 'cross-border dwelling', a complement of cross-border working.

The first part begins with a markedly theoretical chapter that is based on the results of empirical studies in different German borderlands. The author, sociologist Nils **Müller**, offers a good entry into the overall issue by questioning the general assumption that cross-border activities in borderlands tend to increase after the formal opening of the national borders. He argues that each border opening has to be considered as a specific event embedded in the local spatial structure, and proposes a theoretical model that helps to understand how the modifications of rules and regulations concerning local spatial structures, resulting from this event, might affect the beliefs and opportunities of the people living close to the border and thus lead to changes in their routines – changes, however, that are restricted to specific areas of everyday life and vary between individuals as well as between border regions. The model is applied to data on the changing recreational activities in two border villages located at the German-Dutch and German-Swiss borders.

The following chapter on the commemoration of a historical event in the Danish-German border region likewise questions the idea that a formal border opening necessarily leads to increased cross-border interaction. Using the example of a recent Interreg project (Dybbøl 2014) that was intended to foster cross-border integration via recourse to history or cultural heritage, the historian Martin **Klatt** demonstrates the inappropriateness of EU policymaking in border regions and argues that its relative failure is due, among other factors, to not taking into account the composition and diversity of borderland populations. In the Danish-German case where the current politico-territorial boundary does not coincide with ethnico-linguistic borders, this composition is particularly complex. Klatt proposes to analyse this complexity and to consider the factors that encourage or discourage individual cross-border mobility by using a theoretical framework that relies on the notion of (un)familiarity and distinguishes four categories of borderlanders (nationals, transnationals, border surfers and regionauts).

The last chapter in the first part is by Jarosław **Jańczak**, a political scientist who presents yet another way of approaching the surprising fact of limited cross-border activity – or, as the author terms it, of social (re-)frontierization in a period of spatial deboundarization. With the example discussed by him, the Polish-German borderland, we not only leave the narrow realm of Western European borderlands but are also confronted with the inadequacy of the idea of borderland as a continuous entity constituted by the boundary between two countries. Focusing on several Polish-German border (twin) towns that are seen as 'laboratories of European integration', Jańczak presents us with a highly diversified borderscape. The examples of Löcknitz-Szczecin, Frankfurt (Oder)-Słubice and Görlitz-Zgorzelec represent three different patterns of cross-border suburbanization – described by the author as 'colonization', 'cohabitation' and 'reversed suburbanization' – that have distinct consequences for local polity creation and cross-border politics.

Following these different and in part complementary analytical models that help us understand everyday processes in borderlands, the contributions presented in the second part of the volume are dedicated to diverse social entities – the village and the city, families and individual members of a borderland minority – and the question of how these are shaped by the border. The first chapter is about a rapidly changing border village in the Greater Region SaarLorLux in which ethnographic research on residential mobility across the Luxembourgian-German border was conducted. Cross-border residential mobility – that is, the process by which people relocate their residence a short distance across a national border – is a steadily growing phenomenon in the European Union, including the borderlands surrounding Luxembourg. It is in this context that Gregor **Schnuer** explores different notions of home and their connection with borders and border crossings. Unlike most of the work on borderlands, his chapter is not on mobile borderlanders but instead looks at the autochthonous population of a German border village, or more specifically at two representatives of this population. Schnuer argues that the diverging perceptions of the defining features of their village, and how these relate to the large influx of new residents, show the ways in which cross-border activities, especially residential mobility, impact the identity of the residents as members of a particular community. The significant number of new residents does not play out in a national dichotomy of 'us Germans' and 'they Luxembourgers' or generate a sense of local unity among autochthons. Instead, so it is argued, it presents the autochthons with an unusual circumstance that brings differing opinions about the local identity to the foreground.

The next chapter deals with a very different border location, or more precisely with a particular moment in a location's history, namely Baden-Baden's short-term rise to become the 'queen of watering places' in the mid-nineteenth century, a rise that was due in part to its proximity to the French-German border and which came to a complete halt with the advent of the Franco-Prussian War of 1870–71. In her chapter Eva **Zimmermann** presents the cosmopolitan world of Baden-Baden from an unexpected angle by focusing not on the international upper class guests but on the cross-border seasonal workers and foreign merchants and their relationships to the autochthonous population. The author argues that these working and lower middle class people were more intensively engaged in transnational interactions than the members of the leisure class.

While Zimmermann gives a vivid account of a case of rapid and far-reaching de-bordering – in the course of three decades, Baden-Baden had developed into 'Europe's summer capital' – and the subsequent rapid re-bordering, the next chapter is about more complicated, conflicting and ongoing border processes. By analysing three novels by contemporary Belgian writers who belong to the country's German-speaking minority, Lesley **Penné** acquaints us with the identity problems that stem from shifting national boundaries and the ensuing unresolved memories. Although the authors in question develop differing notions of borders and border-related identification processes, they share, according to Penné, the view that it is exactly the fact of living in a – variously conceived – borderland that enables people to reconcile with the past and with one's *Heimat*.

The last chapter in the second part again leaves the narrow circle of Western Europe. Keiu **Telve** presents the first results of her anthropological research on Estonian commuters to Finland. In contrast to the previous case of cross-border labour mobility in nineteenth century Baden-Baden, we are here confronted with a contemporary phenomenon, the bordering effects of which are difficult to assess in their entirety. The chapter focuses on the impact of border crossing on family relations. Starting from the observation that positive aspects of commuting tend to be largely neglected in research – a variant of the aforementioned propensity of border studies to concentrate on crisis – Telve gives a more comprehensive view, including family-related reasons for commuting and the effects of commuting that enhance family life and marital relations rather than disturb them. By showing the interrelatedness of practices of mobility and border crossing with other interactions – such as family relationships – and by also considering the attitudes and experiences of those who stay, Telve arrives at a fuller rendering of this aspect of everyday life in borderlands.

The last part of the book is dedicated to a specific form of borderland mobility: the residential move to the other side of the border. This movement, which some consider to be migration and others do not, is becoming more and more frequent at several inner European borders, among them those within the Greater Region SaarLorLux (see also the chapter by Gregor Schnuer). The first contribution to this part is by Birte **Nienaber** and Isabelle **Pigeron-Piroth**, who provide us with quantitative data on the development of this phenomenon in the Greater Region over the last twenty years. The analysis of these data, which mainly consists of a comparison of distinct three-year periods, reveals how difficult it is to arrive at a general assessment of the demographic developments and their impact on bordering processes. These findings are coupled with qualitative results that show that the motivations for border-crossing residential movement are not constant, but depend upon, among other things, the development of the family.

The chapter that follows also focuses on the Greater Region and brings us back to the notion of (un)familiarity that played a role in Klatt's work on the German-Danish borderland. This notion has gained some prominence in border research lately, and in research on cross-border consumer practices in particular (Spierings and van der Velde, 2013). Elisabeth **Boesen** and Gregor **Schnuer** explore whether the concept of (un)familiarity proves fruitful to the analysis of moving home and dwelling across the border, and, if so, whether the way (un)familiarity features here might inform further conceptual work. By analysing different groups of residential migrants in German border villages, they critically assess the (implicit) equation of familiarity/unfamiliarity with sameness/difference and try to show that this confusion might be caused, at least in part, by the neglect of the relevance of proximity, of being in the same place, to feelings of (un)familiarity.

A further chapter on cross-border residential mobility deals with recent developments at the Belgian-French border. Garance **Clément** discusses the results of her research on the agglomeration of Lille, a typical example of this rather new form of borderland mobility which develops primarily in the proximity of urban centres and can thus in most cases be seen as being part of a more general process of sub-urbanisation (Jagodic, 2012). Her way of looking at these developments is, however, far from typical. Clément starts from the hypothesis that border transgression is experienced by members of the French middle classes as a process of upward social mobility. Relying on qualitative interview data, she shows that this social upgrading is in turn dependent on 'border socialization' and specific resources that support migration.

The last chapter in the book presents another example of an eastern European borderland. Tamàs **Hardi** investigates cross-border interaction – residential movement being just one form – in the Hungarian-Slovak borderland. Like Jańczak in his study of the German-Polish border, he is confronted with marked diversity. Leaning on a dynamic interpretation of Martinez's framework of types of borderland interaction and borderlanders and using data from several quantitative studies, Hardi not only arrives at describing the area along the Hungarian-Slovak border as being composed of different borderlands, but also demonstrates that being a transnational borderlander can rely on very diverse attitudes and resources.

In conclusion, we might ask whether the diversity observable along the two eastern European borders is revealing with regard to the difference between established and relatively new borders and the research preferences that go with it. In spite of the much-discussed idea of regional integration and the coming 'Europe of regions', the solidifying of borders – including their gradual opening – seems to be accompanied by homogenization in the sense that national dichotomies increasingly replace or veil the disparities which exist on both sides, and can, as is shown by the two cases presented here, at least initially be furthered by cross-border interaction. These contrasting cases thus imply a plea for comparative and differently scaled studies along one and the same border, ranging from regions and cities to individual residential areas and, ultimately, to close-ups which focus on families and individuals.

Notes

1 'Greater Region SaarLorLux' also refers to a 'Euroregion' that has existed since 1995; for further information on the Greater Region, see Wille, 2015.
2 In this context Spierings and van der Velde (2008) speak about a 'border-paradox', 'where increasing cross-border integration ... could coincide with increasing cross-border immobility' (2008, p.503); see also van Houtum and van der Velde, 2004, and Müller in this volume.
3 With the exception of the article by Boesen and Schnuer, all contributions in this volume were initially developed as papers presented at a workshop held at the University of Luxembourg in November 2014. We would like to thank the Fonds National de la Recherche Luxembourg for the financial support for this meeting.
4 See e.g. Rumford's (2008) notion of 'borderwork'.
5 Among others, see O'Dowd (2001) on trans-frontier regionalism as an indicator of the way borders are being reconfigured in the EU.
6 A small selection of recent edited volumes reflects the diversity of these perspectives: eastern European borders (Meinhof, 2002); southeastern European versus northeastern European borders (Lauth-Bacas and Kavanagh, 2013); eastern Europe and America (Pavlakovich-Kochi et al., 2004); the global scale (Johnson and Jones, 2014; Brambilla et al., 2015); and the 'established Europe of the West' and the 'emerging Europe of the East' (Kaplan and Häkli, 2002).

Bibliography

Balibar, E., 2003. *We, the People of Europe. Reflections on Transnational Citizenship*. Princeton: Princeton University Press.

Barth, F., 1969. Introduction. In F. Barth, ed., 1969. *Ethnic Groups and Boundaries: The Social Organization of Cultural Difference*. Oslo: Universitetsforlaget, pp.9–38.

Boesen, E. and Schnuer, G., 2015. Wohnen jenseits der Grenze: Regionale Integration und ihre lokale Verwirklichung. In C. Wille, ed., 2015. *Lebenswirklichkeiten und politische Konstruktionen in Grenzregionen: Das Beispiel der Großregion SaarLorLux: Wirtschaft – Politik – Alltag – Kultur*. Bielefeld: Transcript, pp.179–201.

Brambilla, C., Laine, J., Scott, J. W. and Bocchi, G., eds., 2015. *Borderscaping: Imaginations and Practices of Border Making*. Farnham: Ashgate.

Häkli, J. and Kaplan, D. H., 2002. Learning from Europe? Borderlands in Social and Geographical Context. In D. H. Kaplan and J. Häkli, eds., 2002. *Boundaries and Places: European Borderlands in Geographical Context*. Oxford: Rowman & Littlefield Publishers, pp.1–17.

Jagodic, D., 2012. Living (beyond) the Border: European Integration Processes and Cross-border Residential Mobility in the Italian-Slovenian Border Area. In D. Jagetic Andersen, M. Klatt and M. Sandberg, eds., 2012. *The Border Multiple: The Practicing of Borders between Public Policy and Everyday Life*. Farnham: Ashgate, pp.201–218.

Johnson, C. and Jones, R., eds., 2014. *Placing the Border in Everyday Life*. Farnham: Ashgate.

Kaplan, D. H. and Häkli, J., eds., 2002. *Boundaries and Places: European Borderlands in Geographical Context*. Oxford: Rowman & Littlefield Publishers.

Lauth-Bacas, J. and Kavanagh, W., eds., 2013. *Border Encounters. Asymmetry and Proximity at Europe's Frontiers*. Oxford: Berghahn Books.

Löfgren, O., 1999. Crossing Borders: The Nationalization of Anxiety. *Ethnologia Scandinavica*, 29, pp.5–27.

Martinez, O., 1994. The Dynamics of Border Interaction: New Approaches to Border Analysis. In C. H. Schofield, ed., 1994. *World Boundaries: Global Boundaries Volume I*. London: Routledge, pp.1–15.

Meinhof, U. H., ed., 2002. *Living (with) Borders. Identity Discourses on East-West Borders in Europe*. Aldershot: Ashgate.

Minghi, J. V., 1991. From Conflict to Harmony in Border Landscapes. In D. Rumley and J. V. Minghi, eds., 1991. *The Geography of Border Landscapes*. London: Routledge, pp.15–30.

Newman, P., 2011. Contemporary Research Agendas in Border Studies: An Overview. In D. Wastl-Water, ed., 2011. *The Ashgate Research Companion to Border Studies*. Farnham: Ashgate, pp.33–47.

Newman, P. and Paasi, A., 1998. Fences and neighbours in the postmodern world: Boundary narratives in political geography. *Progress in Human Geography*, 22(2), pp.186–207.

O'Dowd, L., 2001. State borders, border regions and the construction of European identity. In M. Kohli and M. Novak, eds., 2001. *Will Europe Work?* London: Routledge, pp.95–110.

O'Dowd, L., 2002. The Changing Significance of European Borders. *Regional and Federal Studies*, 12(4), pp.13–36.

Paasi, A., 2011. A Border Theory: An Unattainable Dream or a Realistic Aim for Border Scholars? In D. Wastl-Walter, ed., 2011. *The Ashgate Research Companion to Border Studies*. Farnham: Ashgate, pp.11–31.

Paasi, A., 2014. The Shifting Landscape of Border Studies and the Challenge of Relational Thinking. In M. Bufon, J. Minghi and A. Paasi, eds., 2014. *The New European Frontiers*. Newcastle: Cambridge Scholars, pp.361–377.

Parker, N., Vaughan-Williams, N. et al., 2009. Lines in the Sand? Towards an Agenda for Critical Border Studies. *Geopolitics*, 14, pp.582–587.

Parker, N. and Vaughan-Williams, N., 2012. Critical Border Studies. Broadening and Deepening the 'Lines in the Sand' Agenda. *Geopolitics*, 17, pp.727–33.

Pavlakovich-Kochi, V., Morehouse, B. J. and Wastl-Walter, D., eds., 2015. *Challenged Borderlands. Transcending Political and Cultural Boundaries*. Farnham: Ashgate.

Rumford, C., 2008. Introduction: Citizens and Borderwork in Europe. In C. Rumford, ed., 2008, *Citizens and Borderwork in Contemporary Europe*. London: Routledge, pp.1–12.

Rumford, C., 2012. Towards a Multiperspectival Study of Borders. *Geopolitics*, 17, pp.887–902.

Sohn, C., 2015. On Borders' Multiplicity: A Perspective from Assemblage Theory. *EUBORDERSCAPES Working Paper*, 10.

Spierings, B. and van der Velde, M., 2008. Shopping, Borders and Unfamiliarity: Consumer Mobility in Europe. *Tijdschrift voor Economische en Sociale Geografie*, 99(4), pp.497–505.

van Houtum, H., Kramsch, O. and Zierhofer, W., eds., 2005. *B/ordering Space*. Aldershot: Ashgate.

van Houtum, H. and van der Velde, M., 2004. The Power of Cross-border Labour Market Immobility. *Tijdschrift voor Economische en Sociale Geografie*, 95(1), pp.100–107.

van Houtum, H. and van Naerssen, T., 2002. Bordering, Ordering, Othering. *Tijdschrift voor Economische en Sociale Geographie*, 93(2), pp.125–136.

van der Velde, M. and van Houtum, H., 2004. The Threshold of Indifference. Rethinking Immobility in Explaining Cross-border Labour Mobility. *Jahrbuch der Regionalwissenschaft*, 24, pp.39–49.

van der Velde, M. and van Naerssen, T., 2011. People, Borders, Trajectories: An Approach to Cross-border Mobility and Immobility in and to the European Union. *Area*, 43(2), pp.218–224.

Wille, C., 2012. *Grenzgänger und Räume der Grenze: Raumkonstruktionen in der Großregion SaarLorLux*. Frankfurt am Main: Peter Lang Verlag.

Wille, C., ed., 2015. *Lebenswirklichkeiten und politische Konstruktionen in Grenzregionen: Das Beispiel der Großregion SaarLorLux: Wirtschaft – Politik – Alltag – Kultur*. Bielefeld: Transcript.

Wilson, T. M. and Donnan, H., 1998. Nation, State and Identity at International Borders. In T. M. Wilson and H. Donnan, eds., 1998. *Border Identities. Nation and State at International Frontiers*. Cambridge: Cambridge University Press, pp.1–30.

Part I

Border crossings and border politics

1 A routine-based model of everyday mobility in border regions

Nils Müller

In the last decade, research on national borders has moved from being on the fringes of the social sciences to being one of the most insightful fields of research pertaining to what might be the most consequential development at the end of the twentieth and the beginning of the twenty-first centuries: the dissipation of the nation state. While large-scale processes of globalization, as well as the global mobility of people, goods and capital, have been at the centre of discussion for decades, smaller-scale developments like everyday cross-border mobility have been comparatively neglected. As most people spend much of their time in close proximity to their place of residence (Ellegard and Vilhelmson, 2004), everyday transnational mobility can best be analysed in a context where it is not necessarily connected with longer travel time: in border regions.

More specifically, this chapter aims at developing a better understanding of everyday cross-border mobility in the context of European border regions, where border controls were abolished in the 1990s and where general cultural proximity and strong economic interaction lead us to expect an emergence of firmly established routines of everyday cross-border mobility. Empirical studies over the last decade and more show that those expectations have not yet been realized. While most researchers consider this a surprise and a research object worthy of attention, this chapter will provide a different approach, outlining a routine-based model of everyday mobility in border regions that considers cross-border 'immobility' to actually be the default. The first section argues that local everyday cross-border mobility is, at best, loosely connected to global processes of transnationalization and also quite unaffected by political and administrative attempts to establish cross-border regions. The second and the third sections then outline two theoretical concepts – routines and space – that provide the foundation of the routine-based model of everyday mobility in border regions presented in section four. Section five illustrates the usefulness of this model by looking at routines of recreation in two German border regions.

Everyday mobility in border regions

Before I outline a routine-based model of everyday mobility in (European) border regions, it is important to untangle the relationship between global processes of transnationalization and their expression within specific geographical regions, as well as their manifestation within an individual's everyday life. Most research on transnationalization focuses either on looking at the emergence of a global functional elite or at the establishment of transnational migratory networks. These two strands of research consider very specific groups of people who, for one reason or another, have decided to become transnational agents, and they take for granted the transnational nature of social interaction. In border regions, in contrast, not every person needs to develop intense cross-border routines merely because they live close to a border. This changes the nature of the research profoundly, as it places special emphasis on the question of why people engage in cross-border activities in the first place. Accordingly, everyday mobility in border regions is not just another expression of a general trend of transnationalization; rather, it has to be considered a distinct phenomenon, governed by different dynamics. It is shaped mainly on the local level and is only influenced by global-level processes insofar as those manifest within the local socio-spatial structure.

Since the implementation of the Schengen Treaty and the abolishment of border controls between most of the member states of the European Union, research on the emergence of cross-border activities in everyday life has centred on Europe. The expectation was that the opening of national borders would lead to a surge in cross-border activities, engendering socio-spatially integrated border regions that could be considered as individual nuclei for the social integration of Europe (Blatter, 2004; van Houtum and Boekema, 1994). Still, empirical studies come to the disheartening conclusion that all attempts at creating cross-border 'atmosphere[s], symbols, images, auras, experiences and events' (Löfgren, 2008, p.106) have only brought forth 'brand[s] without meaning' (Matthiessen, 2004, p.38) while people continue to 'glide by one another – exactly as they had done in their contact with each other throughout the twentieth century' (O'Dell, 2003, pp.39–40). Research even shows how closer proximity and stronger interaction might lead to mutual distrust (Boehnke et al., 2007) and fear (Leontidou, Donnan and Afouxenidis, 2005), strengthening processes of 'othering' (Balogh, 2013) and thereby reinforcing mental borders while physical ones disappear. Research on the relevance of the neighbouring country for borderlanders' life planning comes to similar conclusions for the German-Dutch and the German-Danish border regions (de Gijsel and Janssen, 2000; Yndigegn, 2003). Analyses of specific types of activity like shopping or commuting show a more nuanced perspective, according to which border regions display a certain but rather low level of cross-border interaction (Bouwens, 2004; Buch, Schmidt and Niebuhr, 2009; Gottholmseder and Theurl, 2007; Müller, 2014; Roose, 2010).

The question most often derived from those observations is why so few people seemingly make use of the many opportunities now easily within reach. Looking at this question, is it not odd that it sees cross-border 'immobility' as the peculiarity in need of explanation? Is cross-border 'mobility' not what we consider to be the 'new kid on the block', the phenomenon of interest? So why do we not look more closely at how everyday cross-border mobility comes about? More precisely, why do we not ask about the mechanisms of how events like the opening of a national border, induced into the established socio-spatial structure of the border region by political decisions on the transnational level, might have a lasting influence in a way that leads to some kind of significant cross-border interaction (Baud and Schendel, 1997)?

To approach cross-border interaction from this direction, this chapter develops a perspective that puts less emphasis on the border as such and focuses more on the spatial structure of border regions in general and the set of everyday practices that borderlanders are acting out within them. It is important to recognize that the spatial structure close to a national border is not exclusively defined by the border but still contains all the other elements that make up spatial structure across the world. Accordingly, this article interprets borders as one aspect of socio-spatial structure and asks how a change in the nature of the border can impact the everyday routines of the people living in the borderland. To be able to analyse how such changes come about, we need to develop a basic understanding of how spatial structure and the everyday activities of individuals interact.

Conceptualizing change in everyday activities

Looking at the micro-level of an individual's behaviour, research is often based on a model of more or less rational balancing of utility and cost, a succession of separate and mostly conscious decisions (Hindess, 1988). In contrast, the model suggested in this chapter follows Anthony Giddens in considering everyday life as a continuous stream of activity in which distinct acts are singled out only retrospectively. This perspective matches American pragmatism, where 'social [action is viewed] as composed of chains or aggregations of actors confronting problem situations and mobilizing more or less habitual responses' (Gross, 2009, p.368).

When understanding everyday mobility as mainly shaped by habituated routines that are activated to face the repeating challenges of everyday life, the focus of analysis shifts. Individual moments of decision – for example, visiting a specific supermarket or attending an event – can no longer be singled out, which requires the reconsideration of traditional concepts like 'utility' and 'cost'. More importantly, changes in everyday routines can no longer be analysed in isolation but have to be seen within their specific socio-spatial and individual contexts. They follow a different and much more inert dynamic and change, slowly but persistently. Everyday mobility therefore becomes less a quick succession of short-term decisions and more a continuous flow of slowly developing routines that can sustainably resolve the challenges faced in everyday life – more by perpetual trial and error than by rational calculation.

If we consider everyday mobility as a constant flow of habitually enacted routines, the question of how it can change puts the emphasis on the general mechanisms of change in such routines. These routines are relatively stable over time and rarely change in reaction to a conscious and well-calculated rational decision at any one point in time. Instead, they change over time in a process that is much more difficult to quantify: a relearning process – termed 'reflexive monitoring' by Giddens (1984) – in which actors change their routines over time (Ernste, 2010). Within this process of unlearning and relearning we can distinguish two different dimensions: 'opportunities' and 'beliefs' (Hedström, 2005). While 'opportunities' are located outside the actor in the environment and the possibility for certain activities that it provides, 'beliefs' have to be considered as being internalized by the actor and as having their foundations within her convictions and ways of seeing the world. Changes in both of these dimensions can actually engender a change in an individual's everyday routines. We will now take a closer look at the actual mechanisms of such change.

Concerning 'opportunities', there are two fundamental changes that can occur. On the one hand, new opportunities can 'arise' – for example, by the construction of a new shopping centre or the opening of a new concert hall. On the other hand, opportunities can 'disappear' – for example, by the closing of a supermarket or the obstruction of a road. In the latter case, if an actor had not included the disappearing opportunity into her routines the disappearance would not directly affect her everyday activities, as she could continue to do what she did before. However, if an opportunity disappears that she had included into her everyday problem solving, there would be an instant necessity for her to change her routines, as the established ones would no longer be available. In contrast, newly emerging opportunities do not challenge existing routines at all, as they simply get added to the set of already existing but not utilized opportunities.

Whether changes in the 'beliefs' of an actor entail a potential for change in the established everyday routines depends on the interplay of two aspects of this change: on the one hand, we have to see whether beliefs relevant to a possible routine improve or worsen. On the other hand, we need to consider whether the change is related to a routine that is already part of the person's everyday life. Thus, we get four possible configurations, of which two stabilize existing routines while the other two offer a potential for change. If 'beliefs' about some course of action that actors currently have in their portfolio of routines become more negative, these actors might start to question their routines and become a little more open for different activities that they have not yet been pursuing: for example, rumours about active pickpockets might lead an actor to visit different parks, or regular traffic jams might lead them to take a different route to work. The same holds true if an actor's beliefs about a route of action she currently does not have in her portfolio become more positive: a gas station has significantly reduced its prices, or a friend has heartily recommended a restaurant in a nearby city. Even though actors will rarely instantly change their routines, such a change in their 'beliefs' might plant a seed that leads to a shift over time, away from the established routines into a different set of activity patterns.

Now that we have seen how everyday mobility can be conceptualized as the perpetual enactment of established routines in the face of everyday life's challenges, the question of the relevance of the socio-spatial environment of an actor for his or her everyday routines will be discussed in the following section.

Conceptualizing national borders in social space

While space has long been considered a given container of social action, the last decades have seen a shift in our understanding of this very basic phenomenon. Starting with Georg Simmel (1995a; 1995b) but mainly developed by authors like Anthony Giddens (1984) and Henri Lefebvre (1991), sociology has come to understand spatial structure as an artefact and a manifestation of social structure and social practice. Lefebvre's concept presented in his book *The Production of Space* (1991), together with Berger and Luckmann's (1966) idea of the social construction of reality, can help us sketch a simplified model of the process of social practice manifesting in socio-spatial structure: social practices are picked up by 'practitioners of space', like urban planners, architects or politicians, who transform the spatial dimension of these practices into 'representations of space', e.g. maps, administrative districts or access rules, fixing social practice and thereby 'objectifying' it in the sense of Berger and Luckmann (1966). In this way, the meaning ascribed to a certain place or region (Giddens, 1984) is stabilized, decoupling it from actual practice. As soon as those interpretations appear as a social reality of their own, they can develop into even more established 'spaces of representations' that no longer rely on formal objectification but are firmly embedded within people's understanding of their environment – their 'mental maps' (see Stea and Downs, 1970). Those maps no longer have to be actively reproduced but have gained a reality of their own that is rarely challenged within everyday life and which has transformed into 'latent memories' (Kuhn, 1994, p.78, my own translation), which, in turn, shape social life. They become part of the shared symbolic universe and appear self-evident and primordial (Berger and Luckmann, 1966).

Having outlined the transformational mechanism of its reproduction, it is now time to take a closer look at the resulting spatial structure on the macro-level – in a cross-section separated from processes of its reproduction for now – and to present a model allowing us to describe the different dimensions of socio-spatial structure outlined above. Dieter Läpple (1991) proposes such a model by distinguishing four dimensions that closely match the transformational mechanism and equally contribute to the reproduction and stability of spatial structure. The first dimension refers to the 'physical and material substrate' of social space – its topological foundation, its physical infrastructure, the artefacts created and used, special events happening at a defined time at specific places and so on. The second dimension is formed by the 'patterns of (inter-)action' on top of this physical substrate – people interacting with each other in certain ways, patterns of movement within an area and the general practices of interaction with the substrate established within a certain space. This dimension is closely connected to the everyday routines and practices outlined above. The third dimension emphasizes 'rules and regulations', both formal and informal, that shape the practices and activities on the micro-level that make up those 'patterns of (inter-)action' on the macro-level. They provide people with an idea of how they are supposed to behave within a specific spatial context and, at the same time, allow them to base their actions on certain expectations about how other actors within a space are going to behave. The fourth and final dimension refers to 'symbols and signs' attached to a social space in general, but also to the various artefacts, patterns of (inter-)action as well as norms and expectations. This dimension can refer to aspects of territorial identity as well as to symbolic places or rituals. Taking those four dimensions of social space – 'physical substrate', 'patterns of (inter-)action', 'rules and regulations' and 'symbols and signs' – together, it becomes clear why spatial structure, once established, is much less malleable than is often thought from a purely constructivist perspective.

Applying the four dimensions outlined above to national borders as elements of the socio-spatial structure of border regions, we find that they manifest in each of those dimensions on multiple, different levels:

- In terms of 'physical substrate', a national border is manifested in barriers that physically prevent people from crossing the border, like fences, walls and armed border guards as well as specifically defined crossing points, physically manifested in checkpoints and toll bars (Shields, 2006). Additionally there are signs that communicate the end of one national space and the beginning of another, and other physical artefacts charged with symbolic meaning about one or the other nation as well as differences in the national landscape on both sides of the border (Paasi, 2001; Edensor, 2006).
- In terms of 'the patterns of interaction', national borders are expressed in a thinning out of interactions between people to both sides. Additionally, cross-border interaction often requires the use of a foreign language, while institutional contact opportunities are mostly organized nationally (Müller, 2014; Roose, 2010).
- In terms of 'rules and regulations', the act of crossing the border is a formally regulated matter that requires the individual to present the necessary paperwork and possess an adequate status (Cooper and Perkins, 2011; Mau et al., 2012; Müller, 2012). Rules and regulations control which goods can be brought across the border, and might restrict the time spent and activities pursued in the other country. Furthermore, a border-crossing also signifies entering the rule of a different legal order.
- Lastly, in terms of 'symbols and signs', the border provides a powerful anchor for distinguishing '[something] across the border in the THERE among THEM, something that can be neglected [from something] HERE among US' (van der Velde and van Houtum, 2004, p.45). In addition to entering a different legal order, crossing a border is also connected to changing into a different cultural order, which might seem intimidating (Spierings and van der Velde, 2008).

When considering the opening of a national border, effects on all these four dimensions can be completely different: an opening on the level of 'rules and regulations' might even lead to a closing of the border in the mind of the borderlanders (Boehnke et al., 2007; Dürrschmidt, 2002; Dürrschmidt, 2006);

significant cross-border interaction might prevail at a formally closed border (Berg, 2000; Martin, 2008) or even be caused by it (Byrska-Szklarczyk, 2012; Perkmann, 1998; Wust and Zichner, 2010). If we now regard the implementation of the Schengen Treaty as an opening of national borders on the dimension of formal 'rules and regulations', we have to ask the question of how this could lead to a change in people's everyday routines and mobility.

Routine change in European border regions

The core of the conceptual model proposed in this paper lies in the connection between the four dimensions of social space on the macro-level and the two main factors influencing an individual's 'reflexive monitoring' of his or her established everyday routines on the micro-level: 'beliefs' and 'opportunities'.

Among the four dimensions outlined in section three, the most obvious one to have a strong influence on people's 'beliefs' is the dimension of 'symbols and signs', which encompasses established thought patterns and world views that are part of the common reality of the people living in an area (Berger and Luckmann, 1966). They force everyone to relate to them – not necessarily to share or accept them, but at least to take a position in relation to them. As such, interpretations and beliefs are often negotiated in direct interaction between individuals (Blumer, 1969) and are transmitted within a community by word of mouth. Accordingly, 'patterns of (inter-)action' also significantly shape people's 'beliefs'. Finally, some aspects of 'rules and regulations' also have a significant impact on people's 'beliefs': informal rules and expectations (Dahrendorf, 1974) – for example, visiting certain places may be considered inappropriate, or not attending the local town festival could be seen as self-segregation or aloofness. In contrast to her 'beliefs', an actor's 'opportunities' are shaped by a different set of dimensions of spatial structure. Formal 'rules and regulations' have a strong influence on how people can move within a certain area, spanning from access restrictions to opening hours of shops. The second dimension shaping an actor's opportunities is the 'physical substrate', as it constitutes the actual things a person can interact with: the shops one can visit, the roads that are in use, the sights that are available and – with a little stretch of the term physical – the events that are held.

We can now see how the interaction between an 'objectified' spatial structure and people's everyday activities works, and how the 'physical substrate' and formal 'rules and regulations' mainly shape people's 'opportunities', whereas informal 'rules and regulations', 'patterns of (inter-)action' as well as 'signs and symbols' significantly influence the 'beliefs' of the actors. If we now consider the formal opening of the border between many European countries to be an externally induced change in the dimension of formal 'rules and regulations' within the border region, we can start to identify conceptual mechanisms for the emergence of cross-border everyday mobility and routines.

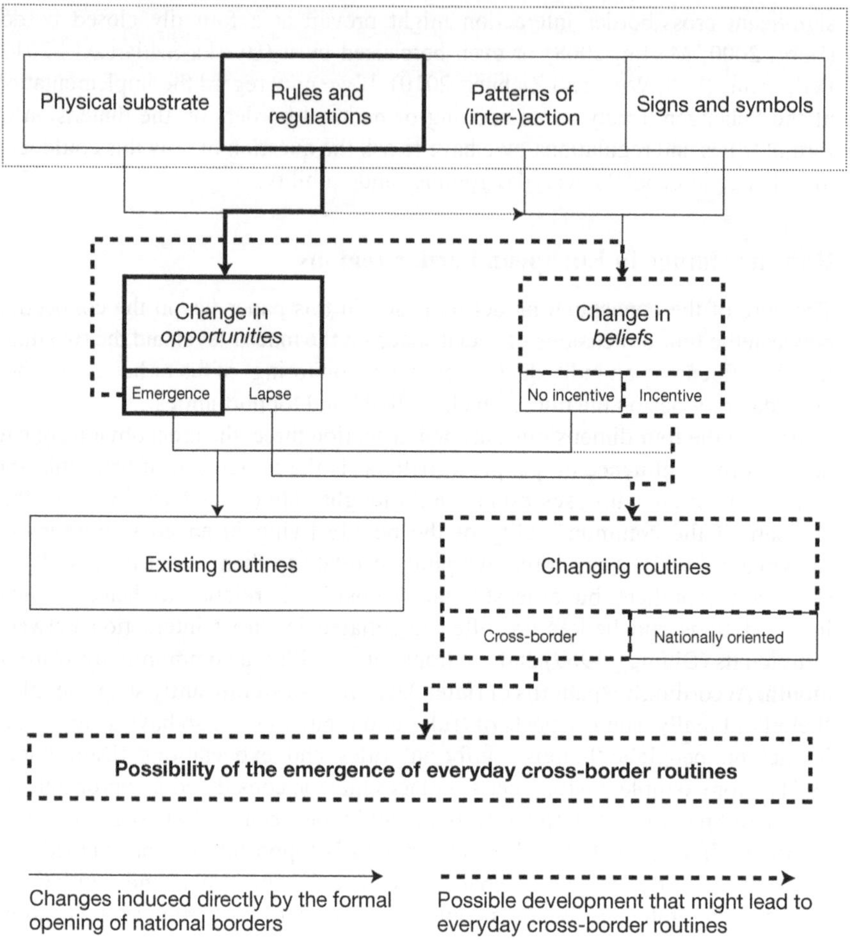

Figure 1.1 Model of the emergence of everyday cross-border routines in border regions.
Source: The author, adapted and translated from Müller (2014).

Taking into account considerations of how the different dimensions of spatial structure affect everyday activities, we see that formal 'rules and regulations' strongly influence the 'opportunities' of an actor. In this case, the formal opening of the border removes specific barriers that made crossing the border difficult, time-consuming and to some extent possibly frightening – effectively preventing options on the other side from being 'opportunities'. After the opening of the border those options suddenly become viable 'opportunities', broadening the portfolio of possible activities and routines. This reasoning brings us to the pivotal point that explains much of the confusion and surprise that can be found in border studies when researchers observe a relatively small amount of cross-border interactions: from the perspective of our theoretical framework, changes in the 'opportunities' of an actor induced by the formal opening of a national border have to be seen as an emergence of new 'opportunities' without any direct influence on everyday routines. The opening of a border does usually not directly influence the routines established by an actor; it does not prevent them from doing what they have always done; and it also does not devalue the 'opportunities' currently used. It does nothing more than add extra 'opportunities' to the already existing set of unused ones.

Accordingly, changes in everyday routines within borderlands have to be mediated by actors' 'beliefs' on different levels. First of all, the newly emerging opportunities on the other side of the border have to surpass what van der Velde and van Houtum (2004) call a 'threshold of indifference': people have to become aware that an 'opportunity' exists and might be relevant for them before it is possible for that opportunity to become part of their everyday routines. In a second step, an actor needs some reason to look more closely at one of those new 'opportunities'. This could be anything from an advertisement to a news report, word of mouth or even plain curiosity. Even then, however, a single use does not constitute a change in an individual's routines. What is needed is rather a continuous utilization that becomes an established part of the actor's everyday life, a process that is as much about establishing new routines as it is about unlearning established ones (Ernste, 2010). This is not meant as an argument that there will be no change within the everyday routines of borderlanders after the opening of the border; rather, the intention is to state that such a change can by no means be taken for granted. It is heavily contingent on specific circumstances, making the borderlanders change their 'beliefs' and then, possibly, deviate step by step from their nationally oriented routines and establish new cross-border ones. This process can be very slow and is full of preconditions and contingencies.

Routines of recreation in border regions

The conceptual model proposed above helps analyse and explain the actual development of cross-border mobility in border regions. It can be used to look into underlying dynamics in different areas of everyday life and even to compare them among different border regions, not by relying on macro variables but by identifying different local manifestations of the same general mechanisms. To illustrate the usefulness of the model and display its strength in recognizing the particularities of a specific empirical case while at the same time being generally applicable, I will now draw upon my own empirical research on everyday routines in European border regions (Müller, 2014). This research is based on eighteen extensive qualitative interviews conducted in two German border regions: the German village of Tüddern, close to the Dutch city of Sittard-Geleen, and the German village of Grenzach, close to the Swiss city of Basel. Those interviews covered everyday routines in three areas of everyday life – 'provisioning,' 'recreation' and 'socializing' (Partzsch, 1964; Partzsch, 1970) – but this chapter will only focus on recreational routines, as they best illustrate the potential for differentiation offered by the model.

In general, we can distinguish two types of recreational or leisure time activities (Stebbins, 2001): on the one hand there is 'casual leisure', which comprises activities that are conducted without much emotional involvement and that function mostly to simply pass the time, relax or socialize. On the other hand, there is 'serious leisure', 'the steady pursuit of an amateur, hobbyist or career volunteer activity that captivates its participants with its complexity and many challenges' (Stebbins, 2001, p.54). As we will see, these two different types of everyday activity are governed by completely different dynamics when it comes to the development and change of routines in border regions. This is mainly due to differences in newly emerging 'opportunities' surpassing the 'threshold of indifference'.

When considering 'casual leisure', the change in the 'opportunities' brought forward by the formal opening of the national border mostly consists in the availability of new infrastructure, like parks, cinemas or event venues. As I argued above, the emergence of new 'opportunities' alone does not automatically induce change into everyday routines. We rather need some kind of trial-and-error process in which borderlanders discover the region across the border and slowly appropriate it into their everyday lives. There are certain characteristics of 'casual leisure' activities that support such a development, as these types of activities are usually short-term and rarely imply long-term commitment. Thus, trying something new does not come at a high price, making the 'threshold of indifference' rather low. Additionally, many such activities require little interaction with people across the border, which makes a language barrier much less of a problem within the 'beliefs' of the borderlanders. In contrast, other activities, like going to the cinema or the theatre, are very much hindered by language differences. In general, 'casual leisure' activities in the two border regions analysed are shaped by 'beliefs' about comfort and reachability. Accordingly, accessibility of the relevant infrastructure is extremely important for the emergence of actual cross-border routines. Research on the German-Swiss border shows how the price of a ticket for public transport or the availability of parking spaces can significantly impede the emergence of cross-border routines in this area (Müller, 2014). As cross-border traffic infrastructure – especially in terms of public transport – is usually rather weak (e.g. Strüver, 2005), in most cases this might be a reason for people to not cross the border regularly.

Everyday routines concerning activities of 'serious leisure', in turn, display completely different dynamics in the two border regions. By definition these activities are not short-term but are characterized by long-term commitments to specific types of activity, be it a certain kind of sport, an interest in a handicraft or a communal activity such as choir singing. The change in 'opportunities' for 'serious leisure' that occurs through a border opening is quite similar to the change in 'casual leisure': new infrastructure becomes available, as do new clubs, new resources, new events and new people interested in the same topic. This does not automatically lead to a change in routines. However, where 'serious leisure' is concerned, people have a much lower 'threshold of indifference' and are likely to try out new things in their area of interest and to actively explore new options, even if doing so requires overcoming difficulties like differences in language (Müller, 2014). In such cases, people have a strong interest and often actively look for ways to deepen their knowledge and find new people to share their passion with. Additionally, they are much more reflective and monitor developments in their field of interest (Stebbins, 2001), which is reflected by some interviewees who have developed a proactive approach to checking out new opportunities within their field of interest on the other side of the border. These characteristics of 'serious leisure' make the emergence of cross-border routines much more likely.

Even so, the potential for these changes is often counteracted by a different type of mechanism: interest in a specific type of 'serious leisure' most often arises during adolescence and is unlikely to change significantly in later years (Nash, 1967; Scott and Willits, 1998). Furthermore, it is, for the most part, highly dependent on the opportunities available in a person's environment and his or her social network at that time, making a perceived lack of 'opportunities' – which would facilitate embracing new 'opportunities' in the neighbouring country – less likely. This is especially true when the type of 'serious leisure' is connected to joining a club or a team, as can be seen in Tüddern where a diverse and lively club environment enables the emergence of a broad variety of interests. However, if there is an established field of 'serious leisure' in which a lack of 'opportunities' is perceived in the actor's own country while new 'opportunities' present themselves in the neighbouring country, these are often included into the everyday routines. This mechanism is seen most prominently in the village of Grenzach, where a sports club started participating in Swiss contests, leading all the active players to cross into Switzerland regularly and making the neighbouring country an unquestioned part of everyday life – especially for youths.

Conclusion

This chapter started from the observation that the formal opening of inner European national borders did not automatically spark a surge of cross-border activity and the development of routines of cross-border mobility in border regions. The chapter then argued that by looking at border openings on the macro level, much of the theoretical discussion about border regions misses an important point: everyday life in border regions is not significantly governed by global trends or political decisions on the national level, but it is shaped by the specific circumstances of a particular region, its socio-spatial structure and the routines already established both individually and collectively. From this perspective, it becomes clear that the observed lack of intense cross-border interaction is not surprising, but rather what is to be expected when approaching the topic with the sociology of space and the theory of social practice in mind.

By focusing on the two central theoretical concepts of everyday routines and social space, this chapter developed a conceptual model to help understand the dynamics prevalent in border regions in the period after the border opens. It combines two central aspects of individual routine change – a person's 'opportunities' and their 'beliefs' – with the four dimensions of socio-spatial structure – 'physical substrate',' patterns of interaction', 'rules and regulations' and 'symbols and signs'. In this way, we have seen that the formal opening of national borders as brought forward by the Schengen Treaty changes the 'rules and regulations' of a border region in a way that does not induce a sudden change in firmly established routines of everyday mobility. Such a change remains possible, but it has to be mediated by a change in peoples' 'beliefs' that are not directly affected by a change in formal 'rules and regulations'. This challenges researchers to examine more closely how everyday routines change in a specific area and also in different realms of everyday life, like 'casual' and 'serious' leisure, for example. Section five demonstrates how such an approach can enrich our understanding of the emergence of cross-border routines. It shows how activities concerning 'casual leisure' follow a completely different dynamic in two German border regions compared to activities concerning 'serious leisure', resulting in different patterns of everyday cross-border mobility. Nevertheless, these two dynamics still can be described using the general model proposed in sections two, three and four.

This chapter shows how using a general model of everyday life in border regions simultaneously helps us to understand one particular case and to develop more general knowledge about the dynamics of border regions and everyday life. It shows how although individual cases always differ on many dimensions, they can be brought together by using a theoretical approach that focuses more on the dynamics underlying social change than the impact of specific variables in a representative number of cases. The chapter will have met its aim if it inspires other researchers to use this approach to complement a more classic, variable or case-centric approach.

Bibliography

Balogh, P., 2013. Sleeping Abroad but Working at Home: Cross-Border Residential Mobility between Transnationalism and (Re)Bordering. *Geografiska Annaler, Series B: Human Geography*, 95(2), pp.189–204.

Baud, M. and van Schendel, W., 1997. Toward a Comparative History of Borderlands. *Journal of World History*, 8(2), pp.211–242.

Berg, E., 2000. Border Crossing in Manifest Perceptions and Actual Needs. In M. van der Velde and H. van Houtum, eds., 2000. *Borders, Regions and People*. London: Pion, pp.151–165.

Berger, P. L. and Luckmann, T., 1966. *The Social Construction of Reality. A Treatise in the Sociology of Knowledge*. Garden City: Anchor Books.

Blatter, J., 2004. From "Spaces of Place" to "Spaces of Flows"? Territorial and Functional Governance in Cross-border Regions in Europe and North America. *International Journal of Urban and Regional Research*, 28(3), pp.530–548.

Blumer, H., 1969. The Methodological Position of Symbolic Interactionism. In H. Blumer, 1969. *Symbolic Interactionism. Perspective and Method.* Englewood Cliffs: Prentice-Hall, pp.1–21.

Boehnke, K., Kindervater, A., Baier, D. and Rippl, S., 2007. Social Change As A Source Of Macrosocial Stress: Does It Enhance Nationalistic Attitudes? *European Societies*, 9(1), pp.65–90.

Bouwens, S., 2004. The Dynamics of Cross-Border Labor: Commuting from the Dutch to the German Part of the Euregio Meuse-Rhine, 1960–2000. *Journal of Borderland Studies*, 19(2), pp.135–153.

Buch, T., Schmidt, T. D. and Niebuhr, A., 2009. Cross-Border Commuting in the Danish-German Border Region – Integration, Institutions and Cross-Border Interaction. *Journal of Borderland Studies*, 24(2), pp.38–54.

Byrska-Szklarczyk, M., 2012. Borders from the Perspective of "Ants": Petty Smugglers from the Polish-Ukrainian Border Crossing in Medyka. In D. Jagetic Andersen, M. Klatt and M. Sandberg, eds., 2012. *The Border Multiple: The Practicing of Borders Between Public Policy and Everyday Life in a Re-scaling Europe*. Farnham: Ashgate, pp.97–118.

Cooper, A. and Perkins, C., 2011. Borders and status-functions: An institutional approach to the study of borders. *European Journal of Social Theory*, 15(1), pp.55–71.

Dahrendorf, R., 1974. *Homo Sociologicus: Ein Versuch zur Geschichte, Bedeutung und Kritik der Kategorie der sozialen Rolle*. Opladen: Westdeutscher Verlag.

de Gijsel, P. and Janssen, M., 2000. Understanding the Dutch-German cross-border labour market: Are highly educated workers unwilling to move? *Tijdschrift voor Economische en Sociale Geografie*, 91(1), pp.61–77.

Dürrschmidt, J., 2002. "They're Worse Off Than Us" – The Social Construction of European Space and Boundaries in the German/Polish Twin-City Guben-Gubin. *Identities: Global Studies in Power and Culture*, 9(2), pp.123–150.

Dürrschmidt, J., 2006. So near yet so far: Blocked networks, global links and multiple exclusion in the German-Polish borderlands. *Global Networks*, 6(3), pp.245–263.

Edensor, T., 2006. Reconsidering National Temporalities: Institutional Times, Everyday Routines, Serial Spaces and Synchronicities. *European Journal of Social Theory*, 9(4), pp.525–545.

Ellegard, K. and Vilhelmson, B., 2004. Home as a Pocket of Local Order: Everyday Activities and the Friction of Distance. *Geografiska Annaler, Series B: Human Geography*, 86(4), pp.281–296.

Ernste, H., 2010. Bottom-Up European Integration: How to Cross the Threshold of Indifference? *Tijdschrift voor economische en sociale geografie*, 101(2), pp.228–235.

Giddens, A., 1984. *The Constitution of Society: Outline of the Theory of Structuration*. Cambridge: Polity Press.

Gottholmseder, G. and Theurl, E., 2007. Determinants of Cross-Border Commuting: Do Cross-Border Commuters within the Household Matter? *Journal of Borderland Studies*, 22(2), pp.97–112.

Gross, N., 2009. A Pragmatist Theory of Social Mechanisms. *American Sociological Review*, 74(3), pp.358–379

Hedström, P., 2005. *Dissecting the Social: On the Principles of Analytical Sociology*. Cambridge: Cambridge University Press.

Hindess, B., 1988. *Choice, Rationality, and Social Theory*. Boston: Unwin Hyman.

Kuhn, N., 1994. *Sozialwissenschaftliche Raumkonzeptionen. Der Beitrag der raumtheoretischen Ansätze in den Theorien von Simmel, Lefebvre und Giddens für eine sozialwissenschaftliche Theoretisierung des Raums*. Dissertation Universität des Saarlandes.

Läpple, D., 1991. Essay über den Raum: Für ein gesellschaftswissenschaftliches Raumkonzept. In H. Häußermann, D. Ipsen and T. Krämer-Badoni, eds., 1991. *Stadt und Raum*. Pfaffenweiler: Centaurus, pp.1–47.

Lefebvre, H., 1991. *The Production of Space*. Oxford, Cambridge: Blackwell.

Leontidou, L., Donnan, H. and Afouxenidis, A., 2005. Exclusion and Difference along the EU Border: Social and Cultural Markers, Spatialities and Mappings. *International Journal of Urban and Regional Research*, 29(2), pp.389–407.

Löfgren, O., 2008. Regionauts: The Transformation of Cross-Border Regions in Scandinavia. *European Urban and Regional Studies*, 15(3), pp.195–209.

Martin, K., 2008. *Das Eigene in der Fremde. Grenzüberschreitende Beziehungen in der deutsch-französischen Kontaktzone des Saar-Mosel-Raumes (1850–1914)*. Ruhr Universität Bochum.

Matthiessen, C. W., 2004. The Øresund Area: Pre- and post-bridge cross-border functional integration: The bi-national regional question. *GeoJournal*, 61(1), pp.31–39.

Mau, S., Brabandt, H., Laube, L. and Roos, C., 2012. *Liberal States and the Freedom of Movement: Selective Borders, Unequal Mobility*. Basingstoke: Palgrave Macmillan.

Müller, A., 2012. Territorial Borders as Institutions. *European Societies*, November, pp.1–20.

Müller, N., 2014. *Die alltägliche Reproduktion nationaler Grenzen*. Konstanz: UVK.

Nash, J. B., 1967. *Philosophy of Recreation and Leisure*. Dubuque: William C. Brown.

O'Dell, T., 2003. Öresund and the Regionauts. *European Studies*, 19, pp.31–53.

Paasi, A., 2001. Europe as a Social Process and Discourse: Considerations of Place, Boundaries and Identity. *European Urban and Regional Studies*, 8(1), pp.7–28.

Partzsch, D., 1970. Daseinsgrundfunktionen. In "Akademie Der Raumfoschung Und Raumplanung", ed., 1970. *Handwörterbuch der Raumforschung und Raumordnung 1 (A-H)*. Hannover: Jänecke, pp.424–430.

Partzsch, D., 1964. Zum Begriff der Funktionsgesellschaft. In *Mitteilungen des deutschen Verbandes für Wohnungswesen, Städtebau und Raumplanung*. Köln, pp.3–10.

Perkmann, M., 1998. Social Integration and System Integration: Reconsidering the Classical Distinction. *Sociology*, 32(3), pp.491–507.

Roose, J., 2010. *Vergesellschaftung an Europas Binnengrenzen: Eine empirisch-vergleichende Untersuchung zu den Voraussetzungengrenzübergreifender Aktivitäten und Perspektiven in der EU*. Wiesbaden: VS Verlag.

Scott, D. and Willits, F. K., 1998. Adolescent and adult leisure patterns: A reassessment. *Journal of Leisure Research*, 30(3), pp.319–330.

Shields, R., 2006. Boundary-Thinking in Theories of the Present: The Virtuality of Reflexive Modernization. *European Journal of Social Theory*, 9(2), pp.223–237.

Simmel, G., 1995a. Soziologie des Raumes. In R. Kramme, A. Rammstedt and O. Rammstedt, eds., 1995. *Georg Simmel. Aufsätze und Abhandlungen. 1901–1908*. Frankfurt/M.: Suhrkamp. pp.132–183.

Simmel, G., 1995b. Über räumliche Projektionen socialer Formen. In R. Kramme, A. Rammstedt and O. Rammstedt, eds., 1995. *Georg Simmel. Aufsätze und Abhandlungen. 1901–1908*. Frankfurt/M.: Suhrkamp. pp.201–220.

Spierings, B. and van der Velde, M., 2008. Shopping, Borders and Unfamiliarity: Consumer Mobility in Europe. *Tijdschrift voor economische en sociale geografie*, 99(4), pp.497–505.

Stea, D. and Downs, R. M., 1970. From the outside looking in at the inside looking out. *Environment and Behavior*, 2(1), pp.3–12.

Stebbins, R. A., 2001. Serious leisure. *Society*, 38(4), pp.53–57.

Strüver, A., 2005. Bor(der)ing Stories: Spaces of Absence along the Dutch-German Border. In H. van Houtum, O. T. Kramsch and W. Zierhofer, eds., 2005. *B/ordering Space*. Aldershot: Ashgate, pp.207–221.

van der Velde, M. and van Houtum, H., 2004. The threshold of indifference: Rethinking immobility in explaining cross-border labour mobility. *Jahrbuch der Regionalwissenschaft*, 24, pp.39–49.

van Houtum, H. and Boekema, F., 1994. Regions Seen as Laboratories for a New Europe: The Applicability of the Flexible Specialisation Model. In J. J. J. van Dijek and J. P. M. Groenewegen, eds., 1994. *Changing Business Systems in Europe, an Institutional Approach*. Brussels: VUB Press, pp.85–109.

Wust, A. and Zichner, H., 2010. "Here is the wall!" – Is it? *Revue d'études comparatives Est-Ouest*, 41(4), pp.171–193.

Yndigegn, C., 2003. Life Planning in the Periphery: Life Chances and Life Perspectives for Young People in the Danish-German Border Region. *Young*, 11(3), pp.235–251.

2 Dybbøl 2014

Constructing familiarity by remembrance?

Martin Klatt

In 2014 the Danish-German border region launched the large-scale Interreg project 'Dybbøl 2014' (Dybbøl 2014, 2014a), intending to use the 150th anniversary of the Danish-German battle of Dybbøl to promote cross-border integration and demonstrate that Germans and Danes now cooperate on many different levels instead of fighting each other. The idea of the project was to connect cultural heritage with regional development and economic growth (Dybbøl 2014, 2014a):

> Die neue deutsch-dänische Geschichte hat bereits begonnen und wird 2014 ihren ersten Höhepunkt erleben, wenn die deutsch-dänische Jugend, Kultur und Wirtschaft sich grenzüberschreitend begegnen, um die gegenwärtige Zusammenarbeit zu feiern – und sie für die Zukunft weiter auszubauen. Denn nur zusammen können wir unser gemeinsames Ziel erreichen: unsere Grenzregion zu einer Wachstumsregion werden zu lassen.[1]

The battle of Dybbøl in the Danish-German War of 1864 resulted in the Danish loss of the duchies of Schleswig and Holstein, and it has long been a Danish national trauma. Historical narratives have sanctified the sacrifice of young Danes and emphasized the deceit experienced at the hands of Denmark's European allies, as well as the evil cunning of Bismarck attacking 'little' Denmark. These narratives have only recently been questioned by the public, partly as the result of the publication of new popular history books (Buk-Swienty, 2008; 2011). For Germany (Prussia), it was the first victory in a series of wars leading to the unification of most of Germany under Prussian rule in the *Kaiserreich* of 1871. Danish-German foreign relations remained rather uneven until Danish membership of the European Community in 1973, if not longer (Klatt, 2006b; Frandsen, 1994). Nevertheless, today a new narrative of cooperation and cross-border integration dominates politics and the media on both sides of the border, though it is unclear just how far this has disseminated into the population (Klatt, 2006a; 2006b). In fact, in spite of a few success stories, cooperation across the border is not uniformly easy, and unfamiliarity as well as indifference characterize the attitude of the majority of borderlanders to the 'other side' (M.L. Andersen, 2004; Klatt, 2006b; 2010; 2012; 2014a; Klatt and Herrmann, 2011; Yndigegn, 2012). Furthermore, the region's cross-border history has not necessarily been an advantage when trying to build political cooperation (Klatt, 2006a).

In this article, I will examine the use of history to promote contemporary agendas of cross-border cooperation, regional development and smart specialization. I will focus on an Interreg project applying the commemoration of a large event, the 150th anniversary of the Battle of Dybbøl in 1864, across a broad spectrum of society, including historical narrative, contemporary people-to-people events, culture and business. The focus is on the Dybbøl 2014 project and its implementation, as well as the degree of its adoption by the population on both sides of the border. On the basis of research carried out for the European Science Foundation's EuroCoreCode project 'Unfamiliarity as Signs of European Times'[2] (Spierings and van der Velde, 2013a; Szytniewski and Spierings, 2014; Klatt, 2014a; D.J. Andersen, 2014), I will focus on contemporary life in the Danish-German border region, using the example of the Dybbøl 2014 project to analyse the appropriateness of EU policymaking in border regions.

The concept of unfamiliarity

The concept of unfamiliarity can be invoked to explain choices of individual mobility across national borders. While it is generally assumed that national borders constrain the flow of products, services, people and capital, the concept explains how differences or otherness (*unfamiliarity*) encourage or discourage cross-border mobility (Baumann, 1995; Spierings and van der Velde, 2008; 2013b; Timothy, 1995). The core of the theoretical framework is the 'Push-Pull model' of migration theory, which provides typologies for the factors promoting and preventing mobility (Lee, 1966).

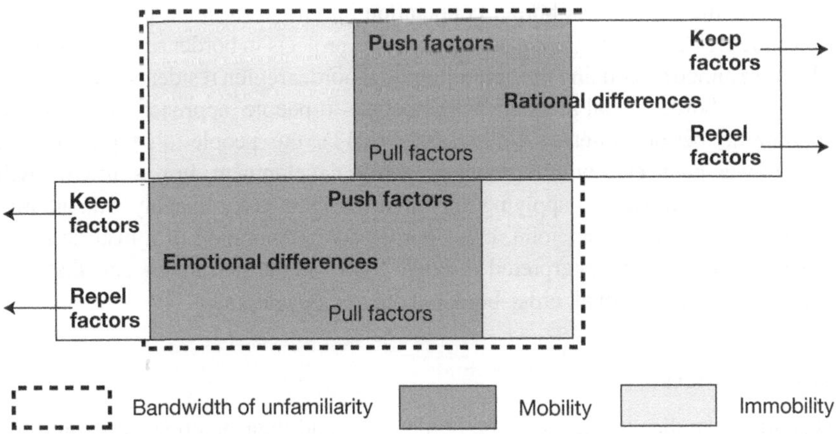

Figure 2.1 Bandwidth of unfamiliarity.
Source: Spierings and van der Velde (2008, p.502).

Applied to consumer mobility, the model has been expanded and adjusted by Bas Spierings and Martin van der Velde (2008). They have added 'keep and repel' factors to the Push-Pull model, as the latter does not explain the lack of mobility within the EU Common Market satisfactorily. Spierings and van der Velde's principal argument is that strong keep and repel factors will result in a high degree of immobility. Unfamiliarity implied by the national socialisation of inhabitants of a border region (Knotter, 2014) would be a strong keep and repel factor, resulting in reluctance to engage in cross-border social practices. On the other hand, unfamiliarity could be an incentive encouraging these cross-border practices, characterized by the adventurous aspect of 'the exotic'. This has been documented historically, most recently for Danish itinerant journeymen at the turn of the twentieth century (Pedersen, 2014), and explains consumer mobility across borders where there is no financial incentive to shop on the other side (Spierings and van der Velde, 2008; also Bygvrå, 2007). Spierings and van der Velde (2008) divide the model into rational and emotional factors that influence the degree of mobility. A rational factor could be an economic situation and an emotional factor could be the feeling of adventurousness. Actual decisions about border crossings are based on a mix of rational and emotional, push-and-pull and keep-and-repel factors. Spierings and van der Velde (2008) define this mixture as the 'bandwidth of unfamiliarity', which is an expression that captures the degree of unfamiliarity one is willing to accept when engaging in cross-border social practices. Cross border practices do not necessarily depend on the reduction of unfamiliarity, as unfamiliarity might as well be an incentive to 'go to the other side'. To determine why interaction happens (or does not happen), it is therefore necessary to define and focus on this bandwidth of unfamiliarity in a borderland's population. In consequence, for policy makers, when more interaction is desired, policies should not necessarily focus on reducing unfamiliarity only, but rather on how to increase the bandwidth of unfamiliarity.

The European Union has acknowledged the issue of unfamiliarity in its border regions by including cross-border cooperation into the Cohesion Policy. Since the 1990s the EU Interreg initiative has co-financed projects in border regions, many of which are aimed at reducing the unfamiliarity of border region residents to encourage more cross-border social practices – the 'people-to-people' approach. In the recent programme period (Interreg IV A, 2007–2013), the people-to-people cultural priority was connected with regional economic development. In this article I will examine this approach by applying the unfamiliarity concept to analyse an attempt to construct familiarity by joint, cross-border commemoration of a historic event that has hitherto been interpreted through controversial narratives, and that was intended to incite regional, cross-border economic development.

Borderlanders

This article studies residents of the borderlands and their practices regarding the border. Borders are not just lines demarcating countries, they both reflect and influence social practices. While borders are social constructions and thus are always open to changes and reinterpretations, social science has been rather uncritical in accepting the bordered territory as a bordered study object, especially when collecting or using empirical data. In consequence, most studies also divide a borderland's population into two national populations, identified by citizenship or place of residence. This causes some problems, as borderland populations in contemporary EU borderlands cannot necessarily be categorized clearly into 'Danes' and 'Germans', for example. For the Danish-German case, where a clear division has been difficult historically as well as in the present, groups within the borderland population have been denoted by the term 'national minority', notwithstanding the lack of clear, objective characteristics to divide a borderland population in general (D. J. Andersen, 2014), but also in particular (Kühl, 2005c; Jebsen and Klatt, 2014). Thus, social science risks falling into the nationalist trap by identifying and categorizing people as nationalized according to the idea of the nation state system, using it as a framework to characterize a population and monopolize, as it were, its identity. As a result, the multiplicity of individuals' spatial identities might not be recognized. The concept of national identity overrides other expressions of identity – localism, regionalism or even national indifference – depending on specific circumstances (Applegate, 1999; Paasi, 2002; Zahra, 2010).

To accommodate this issue, border scholars have introduced the terms 'national' and 'transnational borderlanders' (Martinez, 1994), 'regionauts' (O'Dell, 2003; Löfgren, 2008) and 'border surfers' (Terlouw, 2012) to describe the more complicated reality of borderland populations, which are more difficult to identify individually as well as collectively. All three terms relate to the borderland population's social practices across the border: 'transnational borderlanders' are the opposite of 'national borderlanders'. While the latter are unicultural, often unilingual and rarely cross the border, transnational borderlanders are bicultural, bilingual and have social relations across the border. The term 'regionauts' leaves the area of biculturalism and evaluates border crossing social practices only, focusing on rational choice (i.e. cheaper housing, better job opportunities) and 'exotic' experience, and is thus quite compatible with the concept of unfamiliarity as a motive for the crossing (or not) of a border. Regionauts 'move in both the physical and mental landscapes of a region' (Löfgren, 2008, p.196), exploring the differences and opportunities created by the border. Still, Löfgren (2008) uses the term to describe people identifying with the cross-border region, in this case the Øresund region. To apply the concept to all borderlanders (and border-crossers) profiting from different opportunities on the other side of the border, Terlouw (2012) introduced the term 'border surfers', where border crossing practices are reduced to frequent and temporary affairs that aim to profit from price and cost differentials, based on an intensive knowledge of the societies on each side of the border. Forms of interculturalism do not have to be involved at all.

The aforementioned approaches to understanding borderland populations demonstrate that a comprehensive analysis of the often complicated social practices is most promising when trying to comprehend the complexity of borderland populations. They also demonstrate that familiarity with border-related differences acquired by border crossers, whether they are dubbed transnational borderlanders, regionauts or border surfers, does not necessarily manifest only in crossing the border, but can also result in the perseverance of a mental border, the identification of a matter as 'ours' or 'theirs'. Thus, further observations on the relation between border crossing practices of a borderland's population and the degree or bandwidth of unfamiliarity may lead to new insights about the appropriateness of EU Cohesion Policy approaches and their implementation in border regions.

The Schleswig border region

The Danish-German border region of Schleswig is a classic case of border construction in a border zone applying the nation state's ideal of the highest possible congruency between nation and state. Schleswig belonged to the Danish monarchy, but became closely connected both dynastically and culturally to northern Germany from the Middle Ages onwards. While conglomerate state patriotism in the bourgeoisie can be seen developing from the seventeenth until the turn of the nineteenth century, two national movements claimed the territory for their national projects from the 1840s. Two wars (1848–51, 1864), movement of the border and national assimilation policies followed in the second half of the nineteenth century, climaxing in the 1920 plebiscite, where the region's inhabitants were asked to choose whether they wanted to belong to Germany or Denmark, with no third regional option (Jebsen and Klatt, 2014). In line with the zeitgeist of the 1920s, the dissenters in the plebiscite (about 25 per cent in the northern, now Danish part, and about 20 per cent in the southern, now German part) were accommodated as 'national minorities'. This new phenomenon created organizational confusion in North Schleswig, where the main proponents of organized German-ness had been the Prussian civil servants who had now been transferred elsewhere, and in South Schleswig a small group of very national Danes took on the task of organizing the rather diffuse group of indifferent people, socialists, Frisians and other anti-Prussians, almost exclusively German-speaking, into a 'Danish national minority' (Noack, 1989). By the mid-1920s an institutional framework had integrated these people into a nation state construct (Christiansen, 2008) within the 'host-state', modelled on the kin-state's national institutions. The model was confirmed in 1955 by the Bonn-Copenhagen minority declarations of the Danish and West German governments. It attempted to familiarize the minority populations with their kin-state, not least by employing teachers from or who had at least been educated in the kin-state in the minorities' school systems. Furthermore, Denmark especially has enacted policies to familiarize the North-Schleswigians with the Danish nation state since 1920, denying all forms of local autonomy or special legal status (M. Andersen, 2008) and consequently defamiliarizing Danes and Germans in the border region.

The national dissenters were thus appeased with a generous institutional setup under the framework of minority protection, and figures stabilized at around 12–15,000 German North-Schleswigians and 40–50,000 Danish

South-Schleswigians (estimates based on election results and minority school attendance).[3] While these figures suggest stability, we know that the minorities are in fact very mobile social groups. Identifications depend on concrete life situations and can change during a person's lifetime. Being socially integrated with frequent intermarriages, the specific minority identification can retreat when social practices focus on interaction with majority members, e.g. in the workplace, or alternatively when moving residence to the core nations of Denmark or Germany, often for employment or higher education. For example, the aforementioned telephone poll revealed estimates of about 37,000 people identifying with the Danish minority in Holstein and 25,000 in Hamburg (Schaefer-Rolffs and Schnapp, 2015, p.5). As Schleswig is a peripheral region of both Denmark and Germany, many young people migrate from the region for higher education or other job opportunities, and many families join the minority because of their bilingual kindergarten and school system. In consequence, identification is far from being clear-cut into either Danish or German. Thus, neither the *longue-durée* nor the contemporary identification of the region's population is entirely straightforward, and nor are the terms 'Danes', 'Germans', 'German minority' and 'Danish minority' stable, delimitable or all-inclusive characterizations of the population, either historically or today. Still, the peaceful resolution and reconciliation of a national conflict and a minority settlement are propagated as the political narrative of Schleswig as a region that has transformed from conflict to cooperation, *vom Gegeneinander zum Miteinander und Füreinander* (Klatt, 2015; Kühl, 2005a), assuming a steady reduction of the degree of unfamiliarity in the region.

The shadow of Dybbøl

This narrative of reconciliation and cooperation conflicts with the 'traditional' Danish narrative about Dybbøl, formed shortly after the defeat in 1864. Since then the Dybbøl battle has been a special *lieux de mémoire* for Danes and Danish historiography, while its meaning as a German *lieux de mémoire* has faded over time. It played a role in Prussian-German commemorations during the *Kaiserreich* as the first decisive victory in the wars of unification, but it was later overshadowed by the victory over France in 1871 and the two World Wars. For Denmark, on the contrary, it was the experience of the final deconstruction of its empire after the loss of Scania (1658) and Norway (1814) to Sweden. The loss of Schleswig-Holstein to Germany was the last step that reduced Denmark from the status of an important European power to a small nation-state. The recovery of Northern Schleswig in 1920 is commemorated as a reunification, illustrated as the return of a stolen child to Mother Denmark. Since then the Battle of Dybbøl has been commemorated annually at the battlefield site with the participation of high-ranking politicians and members of the royal family, and many consider Dybbøl the most important *lieux de mémoire* in Denmark. From an international perspective it may seem astonishing to commemorate a military defeat. Still, this *lieux de mémoire* has been very successful in the social construction of Danish national identity as delimited from the larger German cultural sphere, and in characterizing Denmark as a small nation 'for the people' (Feldbæk, 1992). The borderland location was emphasized as a place where Danish language and culture had to struggle and be protected. The sacrifice of young Danes' lives at Dybbøl was interpreted as a central aspect of this struggle and Mother Denmark's continuing protection of Southern Jutland from German cultural expansion.

In 1964, on the one hundredth anniversary, prime minister Jens Otto Kragh was the first Danish politician to accentuate a more contemporary narrative of European cooperation, acknowledging a democratic West Germany that had become an ally and a full member of the Western world (speech quoted in *Grænsevagten*, 46 (6), 1964, pp.182–184). After this, King Frederik IX actually spoke outside the programme, and maintained the importance of Danish soldiers having given their lives to guard the Danish-ness of Southern Jutland and the continuing responsibility of all Danes to do this (speech quoted in *Jyllands Posten*, 19 April 1964). The newspaper *Flensborg Avis*, as well as the nationalist magazine *Grænsevagten* ('Border Guard'), printed the prime minister's speech without comment, but praised the King for his well-chosen words (*Flensborg Avis*, 20 April 1964; *Grænsevagten*, 46 (6), 1964, p.181). It was another thirty years before the chairman of the German minority organization was invited to be a speaker in 1995. In 2001 a delegation from a still existing Westphalian regiment that fought in the battle on the Prussian side was allowed to participate in the commemoration. In 2014, the 150th anniversary of the battle was celebrated with the prime minister of Schleswig-Holstein as the third main speaker after the Queen and the Danish prime minister, and the battle commemoration was used as a symbol for reconciliation, innovation and growth, but also as an occasion for a renaissance of historicizing, including a ten-hour-long TV series, directed by the internationally known Ole Bornedal, which was the most expensive Danish TV production ever made. The year 1864 was a turning point in terminating a pre-national Danish-German history of cultural affinity and political cooperation, and 2014 was intended to mark the turning point toward a post-national history of cooperation, people-to-people reconciliation and regional, cross-border economic development.

Cross-border cooperation

Regional politicians began to meet regularly across the border in the 1970s and 1980s to discuss cooperation. A joint commission of Danish and German politicians managed to work on the issue of pollution of the Flensborg Fjord water inlet, which is divided by the border (Klatt, 2006b, pp.146–153). In general, it became clear that Danish politicians preferred case-oriented cooperation only, while regional German politicians advocated an institutional approach, suggesting the establishment of a Euroregion or a joint spatial planning commission (Klatt, 2006b). When the county of Southern Jutland finally agreed to the German proposal to establish a Euroregion in 1997, this aroused considerable popular protest, often drawing on historical narratives that looked back toward Dybbøl. The Euroregion was thought to confirm a perceived German expansionism, where the 1920 border had to be maintained as a visible line to protect Denmark (Klatt, 2006b; Damsgaard, 1999). Since the Euroregion's foundation, cross-border cooperation activities have increased in number. Still, it is difficult to assess how far-reaching the cooperation is, how much it has disseminated into the border region's population and how it has changed social practices. Cross-border commuting has increased since the 2000s but remains marginal in relation to the whole region's labour market (Buch et al., 2009; Klatt, 2014a). Data on commuters are difficult to assess, but there appears to be a tendency toward positive experiences among both commuters and employers (Klatt, 2014a). Still, many barriers are identified that hinder cooperation within the labour market and education – mostly based on national law not being applicable on the other side of the border. Lack of knowledge of the neighbour's language makes day-to-day communication difficult. Furthermore, there is a lack of information on technical and social issues regarding the neighbouring country (Barten et al., 2006; Klatt and Herrmann, 2011). Finally, a poll conducted among a thousand residents of the border region in 2012 revealed that 83 per cent of the Danes and 74 per cent of the Germans agreed at least 'somewhat' that regional Germans and Danes have much in common. On the other hand, only 29 per cent of the Danes and 26 per cent of the Germans had family or friends on the other side of the border (Region Syddanmark, 2012, pp.5–7). This indicates that the previous estimate of Danish-German familiarity is in fact based on a high degree of unfamiliarity, as most of the respondents do not have many in-depth social contacts (family and friends) on the other side. The respondents who had family and friends on the other side of the border did express a significantly higher rate of agreement that regional Danes and Germans have much in common than those unfamiliar with the neighbouring country.[4]

Ultimately, the narrative of reconciliation and decreasing unfamiliarity is challenged by the success of the right wing, nationalistic Danish People's Party at both the European elections in 2013 and the June 2015 national parliamentary elections in Denmark. In the latter, they earned their best result in the border municipalities of Aabenraa and Sønderborg (well above 30 per cent), and they were the strongest party in all the Southern Jutland municipalities. The recent EU referendum (3 December 2015, on whether Denmark should join the cooperation on Justice and Home Affairs) showed a higher share of 'no' votes in the border region municipalities than the national result. This was the first time that the border region's voters deviated to the 'no' faction compared to the national results of all Danish EU plebiscites since 1972.

So has the shadow of Dybbøl really been lifted, or are we experiencing a re-bordering expressed by a reduction of the bandwidth of unfamiliarity?

The Dybbøl Project

The Dybbøl Project attempted to connect the 150th anniversary of the battle at Dybbøl/Düppel in April 1864 with contemporary cross-border cooperation. It was funded by the Interreg IV Syddanmark – Schleswig – KERN programme (Dybbøl 2014, 2014a). Region Syddanmark (Region South Denmark) co-financed the activities with 8.17 million DKK (around €1.1 million) (Region Syddanmark, 2014). Project partners were Region Syddanmark, the German state of Schleswig-Holstein, the Chamber of Industry and Commerce in Flensburg and the Southern Jutland Development Council (Udviklingsråd Sønderjylland). It was based on a Danish idea discussed in the Syddansk Dybbøl Forum established by Region Syddanmark in 2011, with members representing business, higher education, civil society and cultural institutions from both sides of the border (Region Syddanmark, 2014). The focus was to create impulses for economic growth through culture (Dybbøl 2014, 2014a). Cultural activities based on 1864 were intended to bind the Danish-German region further, to attract tourists and to show the region as an attractive growth prospect for both businesses and residents. The activities were to be focused on the change of attitude since 1864, especially the role of the minorities for cross-border understanding. Additionally, several sports events were planned to visualize the history of 1864 (Region Syddanmark, 2014).

Activities within the project included the following:

- A bicycle race along the Danish-German border with several border crossings on the way (around 350 participants)
- A half-day seminar for Danes and Germans called 'Dybbøl – what now?' in which the participants discussed novels involving the war of 1864
- 'Paradise hoppers', a three-day summer school for 47 elementary schoolchildren. Under the supervision of artists, the children created different artefacts and developed a role-playing game to promote tolerance and cooperation
- A play on the war acted by eighth graders
- An 1864 event during the Danish autumn school holidays at the Dybbøl battlefield site museum with activities for families with children
- A German-Danish business conference that was intended to demonstrate business opportunities in the Danish-German region, with entrepreneurship and innovation as the key forces to create growth (Region Syddanmark, 2014)
- Another business conference, almost simultaneous, under the militant title 'The Battle' was supported by the project (Dybbøl 2014, 2014b)
- The jUNGzuSAMMEN-project (Young Together), a two-week summer school for young people aged eighteen to twenty-five, predominantly from or residing in the border region, to discuss prospects for development and to encourage the young adults to stay in the region (jUNGzuSAMMEN, 2014a)

The project thus comprised an interesting and innovative attempt to connect peace-building and reconciliation with economic growth and cross-border regional development. It attempted to reduce the border region's population's unfamiliarity, with the final goal of spurring growth and development. The project applied the narrative of '*Vom Gegeneinander zum Miteinander*' – from opposition to togetherness – which until the millennium was used to describe the changing situation of the border region's two minorities and their relation to the majority

(Kühl and Weller, 2005; Kühl, 2005b), to the general attitude of the borderland's population towards each other.

Evaluation

The Dybbøl 2014 project has spurred an impressive array of activities, but has it been a true cross-border initiative reflecting reconciliation, integration and joint strategies to develop the region? How far has it been successful? Has it reduced unfamiliarity, and has it been possible to exploit this reduction with tangible impulses for cross-border economic growth and development?

Even though it was financed as a cross-border Danish-German project, the impact was mostly on the Danish side of the border. This is natural, as the Dybbøl narrative is stronger within Danish civil society. Consequently, all links connected to the project's website refer to Danish websites, or websites related to other Interreg projects in the region (Dybbøl 2014, 2014b). Originally, all the various sub-projects had a cross-border approach by involving partners from both sides of the border. Nevertheless, tangible aims for a sustainable reduction of unfamiliarity in a significant part of the borderland's population are difficult to detect. The single initiative attracting most people was a bike race, where the long-term effect on decreasing unfamiliarity must be considered negligible. Other specific events such as the discussion of novels on the war and the activities for schoolchildren only reached out to a small number of people. The special events at the Dybbøl Museum reached out to a predominantly Danish target group, especially since the highly advertised TV series on the battle's history was broadcast on Danish national television at the same time. In Germany, it was broadcast a year later on the niche TV channel *arte*.

The only initiative with a truly sustainable approach to reducing cross-border unfamiliarity was the Young Together summer school, where Danish and German young people aged eighteen to twenty-five lived together in a boarding school setting and developed ideas for the border region's development from a future generation's perspective. This setting encouraged an expansion of the bandwidth of unfamiliarity through common visions and scenarios for a more integrated cross-border region. Results were presented to regional politicians and published in a brochure (jUNGzuSAMMEN, 2014b). The results document that the participants of the summer school were aware of the main issues that impede development: the peripheral location, the lack of attractive programmes of higher education, the poor connectivity of the region to more central areas especially in public transport, the language barrier. Their suggestions are not revolutionary or particularly unexpected, yet none of them have been implemented a year later. All in all, the participants evaluated the course very positively, even though the arrangers admitted several challenges (jUNGzuSAMMEN, 2014c). A year earlier, a similar event had had to be cancelled because there were too few applicants. It was especially difficult to attract interested young people on the German side of the border (where population density is twice as high). The author actually tried to recruit a young politician from the German side of the border to participate in a panel discussion entitled 'What does Dybbøl mean for me today?' without success, as the criterion was that the person should not have an affiliation with either of the national minorities. This difficulty in finding interested young people might suggest that those who attended should be considered transnational borderlanders. Peter Buhrmann, director of the Højskolen Østersøen and one of the organizers of the Young Together summer school, supports this explanation and mentions that '*wir rennen offene Türen ein*' ('we are breaking down open doors'). Many participants came from the national minorities or had previous experience with cross-border cooperation. Additionally, they were bilingual, making interpreters unnecessary. It was difficult to attract young people from the majority populations, who could be considered the national borderlanders. So while it was encouraging in its approach, the actual effect of the summer school in reducing unfamiliarity has been limited, since it could not reach out to people who were not already acting within the existing bandwidth of unfamiliarity.

Finally, the link between the cultural and historical events and the business and development agendas was not as strong as it could have been. The two business conferences were not connected to the cultural initiatives at all. Neither of the two business conferences resulted in measures to support concrete integration, but instead reflected continuing unfamiliarity as they repeated previous attempts to address the need to remove barriers to cooperation, indicating little progress in this field since a joint business strategy for the borderland was elaborated in 2005 (Schack and Dall Schmidt, 2005).

Coverage in the media was marginal: a search for Dybbøl 2014 (exact words) in the Danish media search engine Infomedia resulted in just six hits for the period of the project (30 September 2013 to 31 December 2014) when searching all national newspapers and the leading local newspaper, *Jydske Vestkysten* in Sønderborg. A similar search revealed that the widely advertised business conference 'The Battle' on 28 March 2014 only received in-depth coverage in the newspaper of the Danish minority, *Flensborg Avis*. The project's business conference on 3 April (Dybbøl 2014, 2014a) received more attention as it included the presentation of the winners of the border region's innovation prize. None of the winners was a cross-border applicant, however – and in fact the prize was split, selecting a Danish as well as a German winner for every category. The selection of the winners had little to do with the cross-border aspect of the innovations (Klatt, 2014b).

Around a year later the Danish and Schleswig-Holstein governments were invited to a joint ministerial conference on growth and development, professionally organized by a Copenhagen area event manager. Here again, Michael Schack and Torben Dall Schmidt, two researchers with a focus on cross-border business development, reflected on the different consultancy reports elaborated on the border region's cross-border development potential in the past ten years, concluding that 'it's time to act' (*Flensborg Avis*, 17 February 2015). Apparently, the necessary information to expand the bandwidth of unfamiliarity has been made available for actors, but it has not led to concrete cross-border activities that have resulted in decisive business cooperation. The ministerial working group installed at the conference did not produce anything other than a joint ministerial declaration of intent to remove barriers for cross-border commuters before the Danish elections in June 2015, which resulted in a change of government. So far, as of Summer 2016, no political measures have been enacted to follow through on the declaration.

Applicability of the unfamiliarity concept

As mentioned above, the unfamiliarity concept has hitherto been applied to consumer behaviour across borders (Spierings and van der Velde, 2008; 2013a; Szytniewski and Spierings, 2014) and commuting (Klatt, 2014a). Now, I will apply it to a broader variety of practices of social interaction across borders by focusing on the Dybbøl 2014 project. The central focus of the analysis is the bandwidth of unfamiliarity, within which cross-border interaction tends to occur. Generalized for cross-border social practices, I define the bandwidth of unfamiliarity as the range where rational differences across the border are significant enough to encourage cross-border interaction, whether in the form of shopping, commuting, housing, leisure activities or even social contacts. At the same time, it is the range where emotional differences (communication, culture, clichés) are not too significant to make social interaction impossible.

In order to assess the bandwidth of unfamiliarity according to this model for the Danish-German border region, it has to be kept in mind that there are few in-depth social contacts (only around 25 per cent have family or friends across the border), there is minimal integration of the labour market, there exist stereotypes about the 'other', and there is a language barrier, at least from Germans toward Denmark. So, are people in the Danish-German border region living together, or are they living apart?

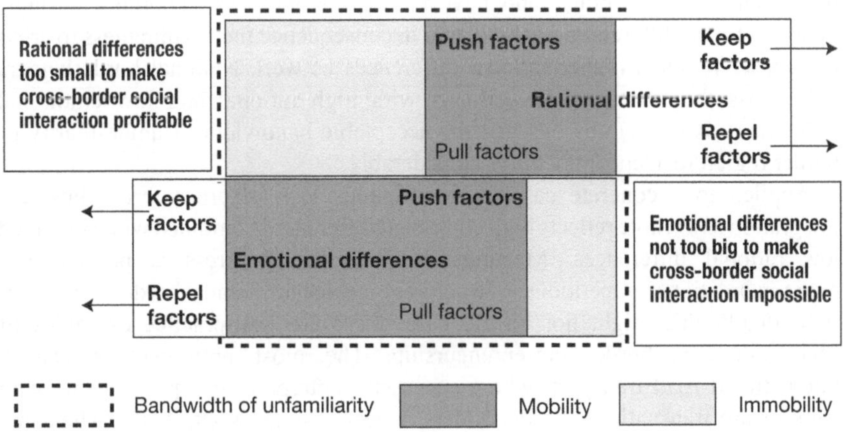

Figure 2.2 Expanded bandwidth of unfamiliarity.
Source: The author, adapted from Spierings and van der Velde (2008, p.502).

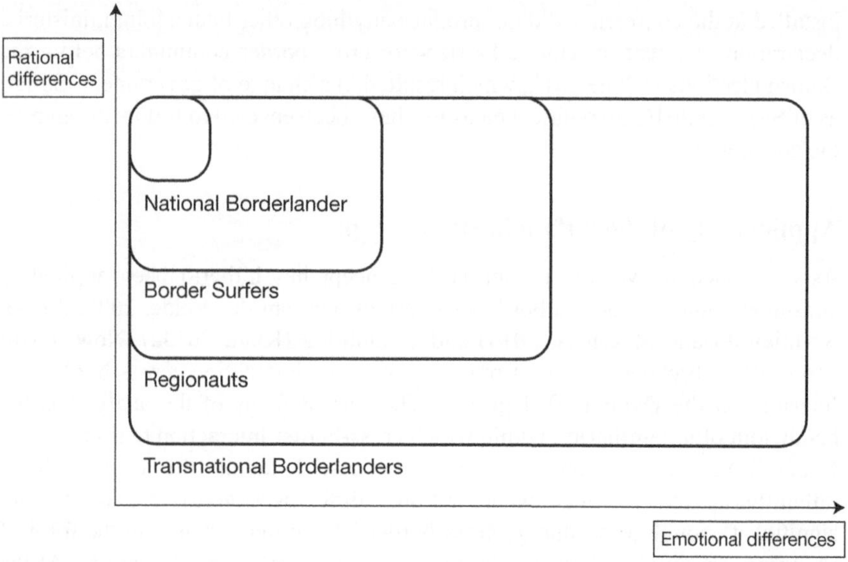

Figure 2.3 Bandwidth of unfamiliarity, types of border region residents.
Source: The author.

In Figures 2.2 and 2.3 I have tried to adapt Spierings and van der Velde's model to social interaction potentials in border regions and classified the individuals who interact according to the categories of transnational borderlander, regionaut and border surfer, as introduced above. Transnational borderlanders' familiarity with the other side results in having low emotional differences regarding the border. They will, and do, cross the border for small rational differences. Their bandwidth of acceptable otherness (unfamiliarity) is thus per se the largest. Regionauts and border surfers, on the other hand, have a more rational focus on border crossings: their emotional differences are high, and in consequence their willingness to cross the border requires higher rational differences as well. National borderlanders only cross the border in borderlands with high rational and low emotional differences. It is also obvious that the acceptable bandwidth of unfamiliarity of border region residents may differ considerably.

Applied to a concrete case, the continuous lack of cross-border business integration as a whole reflects high emotional differences combined with perceived low rational differences. Meetings are frequent, progress is pertinacious, recommendations repetitious, structures persistent, and borders present. Individually this might not be the case: there are best-practice examples of successful cross-border entrepreneurship. The most prominent is Danish international road transport, where successful entrepreneurship by agents who were or are transnational borderlanders is very visible (Klatt, 2013). The same applies to the Young Together project: it was possible to collect a motivated group of young people who were fully aware of the border region's issues. They made sensible suggestions to improve the situation, but the impact was close to zero.

The group was made up of transnational borderlanders, but in the end it did not manage to reach out to the dominant decision-makers, who are national borderlanders.

Conclusion: Dybbøl 2014 – a successful project?

Has the Dybbøl 2014 project reduced the degree of unfamiliarity in the border region by improving conditions for people-to-people interaction and business integration? The cross-border aspect of the Dybbøl project has not been especially visible: most activities happen in Denmark or relate to the importance of Dybbøl for the central Danish historical narrative. The fact that Dybbøl is only of minor importance for the contemporary German and also for the regional Schleswig-Holstein awareness of history remains clearly visible in the activities of the project and the minimal interest it received outside the political circles that were involved. Unfamiliarity regarding the region's common history is still high. Likewise, the connection between commemorating historical events and creating economic growth has not been convincing. The people-to-people activities reached out to Danes and the 'usual suspects', transnational borderlanders with German passports, but very often affiliated with the Danish minority of South Schleswig – that is, people characterized by a low degree of unfamiliarity. It is hard to deny that the aims of the Dybbøl project to use history to expand the bandwidth of unfamiliarity, reunite the border region and create a spill-over effect for regional development were not reached. The project remained traditional, embedded in the Danish historical narratives, and its activities received very little attention in Germany. The regional development aspect was a supplement to a series of similar projects revolving on the same issues without progress on the policy-making side – which, alas, is not characterized by a transnational borderlander agenda and fails to produce tangible policies to extend the bandwidth of unfamiliarity.

Notes

1 'The new German-Danish history has already begun, and it will experience its first climax in 2014 when the German-Danish youth, culture and economy will meet across the border to celebrate the current cooperation – and further increase it for the future. We can only achieve our common goal together: to turn our border region into a region of growth' (author's translation).
2 For further information see www.unfamiliarity.eu
3 As minority membership is not recorded, these methods are the only viable way to assess the size of the minorities (Kühl, 2005c). Recently these figures have been confirmed for South Schleswig by Schaefer-Rolffs and Schnapp (2015, p. 5) in a telephone poll on self-identification.
4 Aggregated SPSS data received from Region South Denmark.

Bibliography

Andersen, D. J., 2014. Do If You Dare: Reflections on (Un)familiarity, Identity-Formation and Ontological Politics. *Journal of Borderlands Studies*, 29(3), pp.1–11.

Andersen, M., 2008. *Den følte grænse. Slesvigs deling og genopbygning 1918–1933.* Aabenraa: Historisk Samfund for Sønderjylland.

Andersen, M. L., 2004. *Grænsen i hverdagen – grænsen i hovedet.* Sønderborg/Aabenraa: Museet på Sønderborg Slot/Institut for Grænseregionsforskning.

Applegate, C., 1999. A Europe of Regions: Reflections on the Historiography of Sub-National Places in Modern Times. *American Historical Review*, 104(4), pp.1157–1182.

Barten, U., Bröcker, J., Herrmann, H. and Klatt, M., 2006. Barrieren und Potentiale der grenzüberschreitenden Zusammenarbeit in der Fehmarnbelt Region. In Institut für Regionalforschung der Universität Kiel, 2006. *Beiträge aus dem Institut für Regionalforschung der Universität Kiel.* Kiel: Institut für Regionalforschung der Universität Kiel, Chapter 5.

Baumann, Z., 1995. *Life in Fragments: Essays in Postmodern Morality.* Oxford: Blackwell.

Becker-Christensen, H., 1990. *Det tyske mindretal i Nordslesvig 1920*–1932. Aabenraa: Institut for Grænseregionsforskning.

Buch, T., Dall Schmidt, T. and Niebuhr, A., 2009. Cross-Border Commuting in the Danish-German Border Region – Integration, Institutions and Cross-Border Interaction. *Journal of Borderlands Studies*, 24(2), pp.38–54.

Buk-Swienty, T., 2008. *Slagtebænk Dybbøl. 18. april 1864.* København: Gyldendal.

Buk-Swienty, T., 2011. *Dommedag Als. 29. juni 1864.* København: Gyldendal.

Bygvrå, S., 2007. *Personal Interaction in European Border Regions.* Aabenraa: Institut for Grænseregionsforskning.

Christiansen, A., 2008. *Die Reproduktion der dänischen Minderheit im 21. Jahrhundert zwischen Kultur, Tradition und Gemeinschaftsglauben.* Tönning/Lübeck/Marburg: Der Andere Verlag.

Damsgaard, J. I., 1999. *Der må være en grænse – osse mellem gode naboer. En analyse af modstanden mod oprettelsen af det grænseoverskridende samarbejdsorgan Euroregion Slesvig.* MA: University of Aarhus.

Dybbøl 2014, 2014a. Om Projektet. Available online at: www.dybboel2014.dk/om-projektet [Accessed 23 June 2015].

Dybbøl 2014, 2014b. Kalender. Available online at: www.dybboel2014.dk/kalender [Accessed 29 July 2015].

Feldbæk, O., ed., 1992. *Dansk Identitetshistorie 3: Folkets Danmark 1848–1940.* Copenhagen: C.A. Reitzels Forlag.

Frandsen, S. B., 1994. *Dänemark – der kleine Nachbar im Norden. Aspekte der deutsch-dänischen Beziehungen im 19. und 20. Jahrhundert.* Darmstadt: Wissenschaftliche Buchgesellschaft.

Jebsen, N. and Klatt, M., 2014. The Negotiation of National and Regional Identity During the Schleswig Plebiscite Following the First World War. *First World War Studies*, 5(2), pp.181–211.

jUNGzuSAMMEN, 2014a. jUNGzuSAMMEN2014. Available online at: www.jungzusammen.dk/jung-zusammen-2014.html [Accessed: 29 July 2015].

jUNGzuSAMMEN, 2014b. Vorschläge des Workshops. Available online at: www.interreg4a.de/dwn505928 [Accessed: 31 July 2015].

jUNGzuSAMMEN, 2014c. jUNGzuSAMMEN2014 – Evaluering. Available online at: www.graenseforeningen.dk/sites/default/files/documents/EvalJungZus_DA.pdf [Accessed: 31 July 2015].

Klatt, M., 2006a. Common, Cross-Border Regional History as an Approach to People-to-People Cooperation and Cross-Border Regional Integration. In M. Hurd, ed., 2006. *Borderland Identities: Territory and Belonging in Central, North and East Europe.* Eslöv: Gondolin, pp.109–146.

Klatt, M., 2006b. *Fra modspil til medspil? Grænseoverskridende samarbejde i Sønderjylland/Schleswig 1945–2005.* Aabenraa: Institut for Grænseregionsforskning.

Klatt, M., 2010. Re-enacting a Region – Why is it so Difficult to Revise Border Changes? In E. Medijainen and O. Merkelsmann, eds., 2010. *Border Changes in 20th Century Europe.* Münster: Lit Verlag, pp.21–30.

Klatt, M., 2012. Mobile Regions: Competitive Regional Concepts (not only) in the Danish-German Border Region. In D. J. Andersen, M. Klatt and M. Sandberg, eds., 2012. *The Border Multiple. The Practice of Borders Between Public Policy and Everyday Life in a Re-Scaling Europe.* Aldershot: Ashgate, pp.55–76.

Klatt, M., 2013. Transportbranchen. En vigtig del af Sønderjyllands erhvervshistorie. In H. S. Hansen, E. Vestergaard and T. A. Vestergaard, eds., *Sønderjysk Erhvervshistorie 1800–2000.* Odense: Syddansk Universitetsforlag, pp.187–206.

Klatt, M., 2014a. (Un)Familiarity? Labor Related Cross-Border Mobility in Sønderjylland/ Schleswig Since Denmark Joined the EC in 1973. *Journal of Borderlands Studies,* 29(3), pp.353–373.

Klatt, M., 2014b. The Danish-German Region's First Innovation Price. How Far Can EU Policies Reverse Peripheral Developments?. *Border Regions in Transition XIV: The Border: A Source of Innovation.* Arras/Lille/Mons, 4–7 November (unpublished conference paper).

Klatt, M., 2015. Von der Abgrenzung zur Grenzüberwindung. Die Minderheiten und die Bonn-Kopenhagener Erklärungen als Wendepunkt der deutsch-dänischen Beziehungen? *Grenzfriedenshefte*, 62, pp.55–64.

Klatt, M. and Herrmann, H., 2011. Half Empty or Half Full? Over 30 Years of Regional Cross-Border Cooperation Within the EU: Experiences at the Dutch-German and Danish-German Border. *Journal of Borderlands Studies*, 26(1), pp.65–87.

Knotter, A., 2014. Perspectives on Cross-Border Labor in Europe: "(Un)familiarity" or "Push-and-Pull"? *Journal of Borderlands Studies*, 29(3), pp.319–326.

Kühl, J., 2005a. Ein europäisches Modell? Die schleswigsche Erfahrung und die nationalen Minderheiten. In J. Kühn and R. Bohn, eds., 2005. *Ein europäisches Modell? Nationale Minderheiten im deutsch-dänischen Grenzland 1945–2005.* Bielefeld: Verlag für Regionalgeschichte. pp.401–707.

Kühl, J., ed., 2005b. *København-Bonn Erklæringerne 1955–2005. De dansk-tyske mindretalserklæringernes baggrund, tilblivelse og virkning.* Aabenraa: Institut for Grænseregionsforskning – Syddansk Universitet.

Kühl, J., 2005c. Nationale Minderheiten im dänisch-deutschen Grenzland – Eine Einführung. In J. Kühl and R. Bohn, eds., 2005. *Ein europäisches Modell? Nationale Minderheiten im deutsch-dänischen Grenzland 1945–2005.* Bielefeld: Verlag für Regionalgeschichte, pp.9–58.

Kühl, J. and Weller, M., eds., 2005. *Minority Policy in Action. The Bonn-Copenhagen Declarations in a European Context 1955–2005.* Aabenraa: Institut for Grænseregionsforskning.

Lee, E. S., 1966. A Theory of Migration. *Demography*, 3(1), pp.115–129.

Löfgren, O., 2008. Regionauts: The Transformation of Cross-Border Regions in Scandinavia. *European Urban and Regional Studies*, 15(3), pp.195–209.

Martinez, O. J., 1994. *Border People: Life and Society in the U.S.-Mexico Borderlands*. Tucson: University of Arizona Press.

Noack, J. P., 1989. *Det danske mindretal i Sydslesvig 1920–1945*. Aabenraa: Institut for Grænseregionsforskning.

O'Dell, T., 2003. Øresund and the Regionauts. *European Studies: A Journal of European Culture, History and Politics*, 19(1), pp.31–53.

Paasi, A., 2002. Bounded Spaces in the Mobile World: Deconstructing 'Regional Identity'. *Tijdschrift voor Economische en Sociale Geografie (Journal of Economic and Social Geography)*, 93(2), pp.137–148.

Pedersen, R. E., 2014. *The Concept of (Un)familiarity and Labour Migration in North Schleswig and Beyond 1870–1920*. Odense: Syddansk Universitetsforlag.

Region Syddanmark, 2012. Det Gode Liv. Regional Udviklingsplan, Grænsepanel. Available online at: http://detgodeliv.regionsyddanmark.dk/publikation/graensepanel/ [Accessed: 20 January 2016].

Region Syddanmark, 2014. Udvalget for dansk-tysk samarbejde, 21 January 2015. Available online at: www.regionsyddanmark.dk/wm460353 [Accessed: 23 June 2015].

Schack, M. and Dall Schmidt, T., 2005. *Grænseoverskridende erhvervsstrategi for Region Sønderjylland-Schleswig*. Aabenraa: Institut for Grænseregionsforskning.

Schaefer-Rolffs, A. and Schnapp, K. U., 2015. *Dänische Minderheit deutlich grösser als bisher angenommen. Ergebnisse der ersten repräsentativen Erhebung zur dänischen Minderheit im Norden Deutschlands*. Hamburg: Universität Hamburg, Fakultät Wirtschafts- und Sozialwissenschaften.

Spierings, B. and van der Velde, M., 2008. Shopping, Borders and Unfamiliarity: Consumer Mobility in Europe. *Journal for Economic and Social Geography*, 99(4), pp.497–505.

Spierings, B. and van der Velde, M., 2013a. Cross-Border Differences and Unfamiliarity: Shopping Mobility in the Dutch-German Rhine-Waal Euroregion. *European Planning Studies*, 21(1), pp.5–23.

Spierings, B. and van der Velde, M., 2013b. Cross-Border Mobility, Unfamiliarity and Development Policy in Europe. *European Planning Studies*, 21(1), pp.1–4.

Szytniewski, B. and Spierings, B., 2014. Encounters with Otherness: Implications of (Un) familiarity for Daily Life in Borderlands. *Journal of Borderlands Studies*, 29(3), pp.339–351.

Terlouw, K., 2012. Border Surfers and Euroregions: Unplanned Cross-Border Behaviour and Planned Territorial Structures of Cross-Border Governance. *Planning Practice and Research*, 27(3), pp.351–366.

Timothy, D. J., 1995. Political Boundaries and Tourism: Borders as Tourist Attractions. *Tourism Management*, 16(7), pp.525–532.

Yndigegn, C., 2012. Reviving Unfamiliarity – The Case of Public Resistance to the Establishment of the Danish-German Euroregion. *European Planning Studies*, 21(1), pp.58–74.

Zahra, T., 2010. Imagined Noncommunities: National Indifference as a Category of Analysis. *Slavic Review*, 69(1), pp.93–119.

3 Cross-border urbanism on the German-Polish border

Between spatial de-boundarization and social (re-)frontierization

Jarosław Jańczak

Introduction

Most European borders have been experiencing a long-lasting and ongoing erosion, which is an element of the continental integration process. However, together with the disappearance of the physical borderlines, the socially constructed limits visible in social, cultural and political behaviour are starting to play an increasingly important role, being at the same time much more difficult to 'neutralize'. Additionally, integration processes at borders are marked by an interesting feature: on the one hand, they belong to traditionally understood national policies, where states negotiate the conditions of mutual relations, balance of power and manifestations of sovereignty. On the other hand, however, the integration project is implemented locally by actors living and experiencing the borders, which channel everyday behaviour in a specific way. In conjunction with these tensions, national policies are often referred to as 'border sensitive', where border actors are described as favouring border openness and liberalization. This vision will be partly undermined in this paper by examining one of the changes to a border resulting from eastern enlargement – cross-border urbanism at the German-Polish border. I analyse a complex set of interrelations there, showing the complicated nature of local answers to the new challenges.

One of the most fiercely debated elements of the European Union's 2004 enlargement in Poland was the threat of a massive influx of Germans, especially in border areas. Surprisingly, however, the reverse phenomenon started to become visible very quickly – Polish migration to the German side, noticeable especially in the areas where urban life is concentrated. This process manifests itself differently at the various parts of the more than 500km-long border. The aim of this paper is to categorize various schemes of cross-border urbanism by examining three places located on the German-Polish border: Löcknitz-Szczecin, Frankfurt (Oder)-Słubice and Görlitz-Zgorzelec, representing accordingly the northern, middle and southern parts of the borderline, but also various combinations of spatial and social variables. They manifest three different patterns of cross-border suburbanization with consequences for polity creation, policy making and local cross-border politics. In the first case, the metropolis located on the Polish side neighbours depopulated rural areas on the German side, which has resulted in significant Polish immigration. On the one hand, this is encouraged by German local authorities; on the other hand, it creates anti-Polish attitudes among the local population. The second case is strongly influenced by the creation of a common cross-border development strategy. In the third, the German side is trying to attract Poles to settle down in carefully renovated Görlitz to avoid a demographic and economic catastrophe. This chapter will investigate how these top-down policies are reflected in the political practices of the Poles and Germans on both sides of the border, examining the extent to which there has been a bottom-up answer following the spatial deboundarization, and whether this has resulted in social (re-)frontierization.

Conceptually, the paper is based on three ideas: first, the concept of border towns as 'laboratories of European integration' (Gasparini, 1999–2000), considering (cross-)border urban structures as spots where wider (continental) processes are concentrated and are much more observable; second, the models of de- and re-bordering, especially the two manifestations of de-boundarization and (re-)frontierization, as phenomena differently reflected at state and social levels (Kristof, 1959); and finally, the concept of cross-border governance and its dimensions (Gualini, 2003), namely political and economic, institutional, symbolic and cognitive, each with corresponding consequences for (cross-)border policy, politics and polity. Methodologically the paper is based on border discourse analysis, statistical analysis (electoral results and public census) and sociological survey analysis (examining the mutual attitudes of Poles and Germans from investigated settlements). It concentrates on the period 2004–2014, investigating a time of intensive border changes in the first decade of the common EU membership of Germany and Poland.

Conceptual framework

The starting point for understating the concept of 'border' can be reduced to the idea of distinguishing 'us' from 'non-us', considering it a tool of inclusion and exclusion. Borders are not static, however, in their location, forms and organization. They change over time, hardening or softening, both as physical barriers in the field but also as social constructs and social representations (Kolossov, 2005, p.610). This changing nature of borders is reflected in academic literature about the idea of bordering (Bürkner, 2015; Newman, 2006; Popescu, 2012; van Houtum and van Naerssen, 2002; Kolossov and Scott, 2013), but also de-bordering and re-bordering (Kolossov and Scott, 2013) as dynamic and repetitive variations of the concept. De-bordering in particular is understood here as lowering the significance of borders, especially due to the European integration process (O'Dowd, 2002) and the EU's economic social and cultural policies (Koeppen, 2015). This changing nature of borders very often creates difficulties and challenges for the actors involved (Newman, 2006). This happens, among other reasons, because borders are products of the past, and changing them is also about changing the past (O'Dowd and Wilson, 2002, p.8). At the same time, de-bordering offers more freedom but collides with the principle of security (O'Dowd and Wilson, 2002, p.24).

I will claim, however, that to understand the changing nature of borders, the categories of de-bordering and re-bordering are too narrow in that they are referring to more or less regular cycles. An important element that is not duly considered when examining these processes is the changing forms of border organization. Consequently, I have proposed a new understanding of change based on the categories of boundary and frontier: boundarization and frontierization (Jańczak, 2013a).

Every border can be organized in at least two ways: as either a 'frontier' or a 'boundary' (Kristof, 1959). The former, dominant in pre-Westphalian Europe and (re)introduced with the European integration process, is a space where the influences, cultures, values and goods of neighbouring political organisms overlap (Walters, 2004, pp.687–688; Browning and Joenniemi, 2008, p.529). It is a zone where two entities or social systems can make contact (Evans and Newnham, 1998, p.185), sometimes a sort of 'no man's land'. At the social level, attitudes of openness dominate, marking at the same time the border as the states' 'beginning' and 'entry', with elements of both sides visible (O'Dowd and Wilson, 2002).

The boundary form of border organization was crucial in the creation of modern states, precisely and linearly setting 'more or less strict territorial limits' (Evans and Newnham, 1998, p.185) and marking exclusive sovereignties (O'Dowd and Wilson, 2002; Germond, 2010). This leads to legal, cultural and even ethnic unification within the boundaries, and alienation from the entities located on the other side. There is not too much room for overlapping structures; uniformity dominates on each border side. The boundary, as opposed to the frontier, is more the 'end of territory' of a state. The changing nature of the construction of borders in Europe is marked by their de- and re-boundarization, as well as de- and re-frontierization. European integration, together with the creation and expansion of the Schengen area, has accelerated this process on two levels: spatial and social.

Having briefly discussed the nature of borders, one can stress that about 10 per cent of Europeans live in settlements located less than 25 kilometres away from a state border (Gasparini, 2008), making borders an element of everyday experience for a significant portion of Europeans. Border towns are consequently described in Europe as forerunners of a continent-wide process, being laboratories of continental integration processes (Gasparini, 1999–2000) where specific phenomena concentrate and are more visible than from the continental perspective (Kaiser and Nikiforova, 2008). This approach down-scales the European project, considering border communities and border towns as laboratories of European integration – spaces distinguished from the surrounding environment, offering relatively controlled conditions, with the possibility of initiating and stopping specific processes (Jańczak, 2013b), being consequently both artificially created and actually existing (Laponce, 1972). Additionally, cross-border cooperation and interactions always contain an element of experimentation (Joenniemi and Sergunin, 2011, p.233). Jouni Häkli (2011, p.21) points out that 'the European Union [...] can be seen as an experimentation [...] allowing for local and transnational ties to bundle and overcome Europe's all too territorial past.'

Permeable borders require a new way of managing interactions. This is usually a matter of cross-border governance. Enrico Gualini (2003) distinguishes three dimensions of this: political and economic, institutional, and symbolic and cognitive. The first concentrates on the changing economies (based on now-common cross-border interests and interdependence) of border territories and 'has emerged [...] as a policy alternative aimed at the reduction of core-periphery disparities through development of the endogenous potential of peripheral regions' (Gualini, 2003, pp.43–50). It results in specific cross-border politics. The second is based on institution building, representing cross-border politics. The last is centred on social mobilization and the construction of cross-border communities based on territorial identities.

These concepts will be used to analyse cross-border urbanism in the German-Polish borderland, testing its level of de-boundarization and frontierization in the context of de-scaled integration. I am consequently interested in the creation of (cross-)border policy, politics and polity, and their diversity along the investigated border. The analysis concentrates on the border towns, seeing them as down-scaled laboratories of integration processes that can reveal significant aspects of de-bordering and re-bordering.

The German-Polish border

It is important to note that the currently observable processes on the German-Polish border are deeply rooted in historical developments. These developments not only shaped the border in the past, but they also still influence the political, economic and social changes occurring there. The location as well as construction of this border is rooted in the Yalta Conference and the political decisions of the three leaders of the Great Coalition. Defeated Germany was to be punished (among other ways, territorially) for its attack on the Soviet Union, which was executed by shifting the Polish state westwards. Consequently, the new border was set along the lines of two rivers, the Oder and the Neisse, and cut off the former German (in historical, ethnic and cultural terms) provinces of Pomerania, Eastern Brandenburg and Silesia (Halicka, 2013). The local population partly fled and was partly expelled; the provinces were transferred to the Polish state and gradually filled with a Polish population. Most of the Poles who moved there originated from two locations: the Polish eastern outskirts annexed by the Soviet Union, where the Polish population had also been expelled or repatriated, and Germany, being mainly forced labourers of the Nazi state. The Polish side of the border was officially labelled 'Regained Territories', stressing the 'original', i.e. historical, Polish character of the provinces. In Germany they were named 'territories under temporary Polish administration', emphasizing the expected restoration of the pre-war territorial order. The new border was formally recognized by East Germany in 1950, by West Germany in 1972 and by the united Germany in 1990 (Halicka, 2013). Throughout the entire Cold War the border remained a highly sensitive issue and a potential source of German revisionism regarding Poland (which kept the latter in a trap of alliance with the Soviet Union). The final border recognition became one of the preconditions of German re-unification, and cross-border cooperation was an essential element of the European integration process. Here, a united Germany could demonstrate its peaceful and reliable face, while Poland could realize its western aspirations.

As the result of the described developments, a new border landscape emerged directly after the Second World War. The western part of the borderland was inhabited almost exclusively by the German population, the eastern by the Polish. The two communities were separated not only by a physical boundary but also by social factors. Linguistic, cultural and ethnic differentiation resulted from lack of contact, but also from missing border experiences on both sides: the German population represented German traditions from the heartland, while the Poles came mainly from the eastern outskirts of pre-war Poland. These structural factors were deepened by decades of separation (with a short break in the 1970s marked by an open border as a display of 'socialist brotherhood') that resulted in disparate and mutually alien political and social structures on both sides of the border.

Both politically and socially, the border represents an example of post-war boundarization, with not only sovereignties exclusively executed on both sides of the border, but also internal homogenization, and consequently national, cultural and economic alienation. The period after the collapse of communism was marked, due to the aforementioned German-Polish reconciliation and European integration process, by de-boundarization tendencies and attempts to 'frontierize' the border. Both the context of Polish-German relations (rooted in the high politics of continental integration) and the bottom-up pressure of local authorities

Map 3.1 The German-Polish border.
Source: The author.

(motivated by de-peripheralization and principles of economic development) were influential as Poland gained entry to the European Union in 2004 and to the Schengen Zone in 2007. The liberal border regime and, finally, the elimination of border controls, changed the human, political and economic landscape of the border. However, this new phase was marked by a set of new challenges, starting with structural issues (deep economic asymmetry between Germany and Poland, marked by a GDP ratio of four to one) and ending with the social alienation of both nations in their mutual relations (with numerous contact points, but mainly limited to the commercial context). All these processes were visibly concentrated in the border areas, especially in the border towns, which played the role of laboratories for relations between Poland and Germany and between Poles and Germans.

The eastern enlargement and border-related accession negotiation problems

The eastern enlargement of the European Union notably involved long-lasting and complicated accession negotiations, framed politically by the Copenhagen Criteria and procedurally by the necessity of accepting the *acquis communautaire*. At least two problems in the negotiations – both among the most important issues in the public debate – were connected to the legacies of German-Polish relations: land acquisition and labour mobility. The former was rooted in historical legacies, the latter in economic asymmetries, and both were expected to cumulate on the border and in the borderland.

Poland was afraid of foreigners coming to Poland and, under the conditions of unlimited flow of capital, buying property there. This argument was presented as a general issue, but was politically motivated. The fear of German citizens buying up large tracts of Polish land, especially in the border areas located close to the German state and border, was supported by two further arguments: first, the historical one, assuming the 'Regained Territories' to be a target of the former inhabitants, who could peacefully and easily buy back their former properties; and second, due to the higher purchasing power of Germans, having on average four times higher personal income and more expensive real estate prices in their country. What later became an argument presented by the anti-European movement, as well as by the negotiators, was the threat of a significant influx of Germans who were expected to colonize and re-Germanize the western outskirts of Poland. Poland therefore proposed transition periods for agricultural land purchase, starting with eighteen years, and obtaining twelve years of limitations. In the case of land leasing the western provinces, which belonged to the German state before the war, received a seven-year transition period of exclusion from purchase or rental, while other Polish regions received only a three-year transition period.

Germany, on the other hand, was afraid of a significant influx of immigrants from the new member states looking for a better future in the West. De facto, due to the long common border, this was a postulate related primarily to Poland. Economic asymmetries forced Germany to ask for a transition period here.

Eventually the formula of 2+3+2 was agreed on, allowing two extensions of the original two-year transition period. Germany used it to the maximum, only eliminating barriers for the unlimited flow of people in 2011.

The eastern enlargement was completed with widening of the Schengen Zone in 2007, eliminating direct checks on the German-Polish border, and distancing the border itself from the functional landscape of a borderland. As expected earlier, this was to strengthen both the processes described above, easing an invasion of German landowners and a Polish labour force. Schengen caused de-bordering, visible in the form of spatial de-boundarization, where the border was no longer easily perceptible. At the same time, social (re-)frontierization was still a project that required further efforts to bring together both groups, facilitate non-commercial cross-border contacts and increase mutual understanding. It is important to note that in the years after the eastern enlargement most of the border-related fears have not yet materialized. Germans have not entered the Polish real estate market to the degree feared. The number of German landowners has been minimal, and, additionally, lags far behind the numbers of Dutch citizens. Poles have started to migrate, but have chosen the United Kingdom as their main destination. Interestingly, after 2011 the direction did not change in favour of Germany. What is more, immigration to Germany is concentrated in non-border areas, occurring rather in the big cities in the middle of the country. At the same time, however, another process has started to be visible: the Polish presence on the German side of the border. However, its nature and character has been much more complicated than a simple migration: it has led to the creation of cross-border urbanism. This phenomenon varies across the different border locations, so to understand it further, three cases have been selected, representing not only three geographical sections of the German-Polish border, but also three different spatial-structural schemes, resulting in three different modes of cross-border urbanism and reactions to it.

The phenomenon of German-Polish cross-border urbanism in three patterns

Before presenting the regional peculiarities, it is relevant to mark the general tendencies common to the entire borderland. First of all, the German side of the border has experienced drastic depopulation after the collapse of communism. This especially affected younger and better-educated Germans, who have moved en masse to Berlin and the western part of the country. Second, among the 600,000 Poles spread around the whole of Germany, only about 10,000 are estimated to live in the border area, with half of these in Mecklenburg-Vorpommern. Most of them represent professional profiles that are missing in the half-empty German border communities; they are economically and socially active, many are well-educated and many are small entrepreneurs. There is also a group of internal 're-emigrants' – people with Polish origins who, after a longer stay in the western part of Germany, decided to stay in the country but move closer to Poland. Some of the Poles commute back to Poland on an everyday basis for professional, educational, cultural and commercial reasons (Frelak and Łada, 2011).

Three reasons for their decision to move to the German side of the border dominate among the declarations of the Polish 'settlers': higher income level, family-friendly social support and prices of real estate that are more attractive than in Poland. However, these features have local peculiarities, which can be seen in the border towns located on various parts of the border. There are three regions on the German side: Mecklenburg-Vorpommern, Brandenburg and Saxony, representing three varied developmental, political and cultural types.

The cases investigated in this paper are: Löcknitz (Mecklenburg-Vorpommern, Germany) and its urban neighbour, the Polish city of Szczecin; Frankfurt (Oder) (Brandenburg, Germany) and neighbouring Słubice (Poland); and Görlitz (Saxony, Germany) together with Zgorzelec (Poland).

The case of Löcknitz (Mecklenburg-Vorpommern) represents a massive influx of Poles to Germany, a phenomenon highly present in both the Polish and German media. The current Polish inhabitants of Löcknitz mainly come from the neighbouring metropolis of Szczecin (which has more than 400,000 inhabitants). It is important to note that the Yalta border divided the province of Pomerania into two parts, leaving its centre, the city of Szczecin, on the Polish side (Białecki, 1992). This meant that it was isolated from the province's hinterland, which remained in Germany and became, especially after the collapse of communism, peripheral, underdeveloped, rural and highly depopulated, with some of the towns and communities losing up to one third of their previous population between 1990 and 2015. Szczecin, on the other hand, became the second most important seaport of Poland, and – despite transformation difficulties – is a vivid metropolis with a shipping industry, an educational sector, tourism and more (Musekamp, 2010).

The situation of Frankfurt (Oder) (Brandenburg) and its twin city Słubice represents the opposite situation with regard to asymmetries. Here, the border shift in 1945 divided a single city and left the historical and industrial centre on the German side. The eastern outskirts soon became the city of Słubice. The same happened with the province: the former Brandenburg lost its eastern regions, which became the peripheral, rural and underdeveloped Polish region of Ziemia Lubuska (Jajeśniak-Quast, 2000). The collapse of communism initiated a depopulation process in Frankfurt (Oder) (Knefelkamp and Griesa, 2003) (the local population decreased in the period 1989 to 2013 from 87,000 to 58,000), whereas the number of inhabitants on the Polish side remained almost unchanged at around 18,000. The weakening demographic potential of Frankfurt (Oder) was balanced by the growing influence of metropolitan Berlin, located only one hour's drive westwards and the destination of everyday commuting for many local dwellers. Słubice, on the other hand, started to benefit from its border location, especially as a 'gateway to Germany', a shopping destination for German customers, the site of investment for mainly western companies and a German-Polish university centre (Musiał-Karg, 2009). The number of Poles permanently residing in Frankfurt (Oder) is relatively low, especially when excluding Polish students who temporarily stay there during their studies at the European University Viadrina (Jańczak, 2007).

Görlitz (Saxony) and its border twin Zgorzelec represent a history very similar to the previous one. The historical centre remained on the German side, while the eastern neighbourhoods beyond the Neisse River became Polish Zgorzelec (Jajeśniak-Quast, 2000). Here, both towns were peripherally located, far from bigger centres (Waack and Haslinger, 2000; Lemper, 2009). On the one hand, in 1989 there began an exodus of Görlitz inhabitants to the west, and the town's population dropped from 74,000 in 1989 to 54,000 in 2013. Zgorzelec today has 32,000 inhabitants, according to official statistics, 4,000 fewer than in the middle of the 1990s. At the same time, especially due to central financial support, the historical centre of Görlitz was carefully renovated, not only bringing back the old splendour to the town but also making it one of the most interesting tourist attractions in the region. Görlitz claims to be the last Silesian town in Germany, referring to a province that is nowadays located in Poland and partly in the Czech Republic, just as Zgorzelec pretends to be the last Polish town of Lusatia, the rest of this originally Slavic region belonging to Germany since the Middle Ages. In both places, however, most of the historical buildings remain empty. Spatial remoteness affects the towns' economic development.

The above insight into the local contexts will aid a more detailed look at the forms and manifestation of cross-border urbanism along the border. With regard to the northern part of the German-Polish border, the economic asymmetries are striking in that they are opposite to the general picture of Germany and Poland. The German side is much more affected by economic and social problems. Unemployment in the Ucker-Randow district reached the level of 19 per cent in 2010, and was 25 per cent in Löcknitz itself in 2008, causing a high outflow of inhabitants and resulting in turn in further depression of the local economy, empty kindergartens, schools and so on, and also a drop in real estate prices and hundreds of empty apartments waiting for residents. On the other side of the border, only 18 kilometres away from Löcknitz lies Szczecin, where real estate is expensive and the city is relatively crowded. For as long as the border was barely permeable, both systems were not only relatively static, but mutually isolated. Liberalization of the border regime resulted in spatial de-boundarization and functionally motivated flows of Poles, buying and renting real estate on the German side of the border where prices were lower than on the Polish side. Many of them still work in Szczecin and commute daily, while others decided to open their own businesses on the German side. This flow has been encouraged by the local authorities, considering the influx from Poland to be a solution for the depopulating local towns and districts, using the existing housing resources and giving impetus to the local economy and society.

Deboundarization resulted consequently in a 'natural process' of cross-border migration (with the local authorities' positive orientation), creating a cross-border metropolis with Szczecin as its centre, and with German satellites being colonized by Polish 'settlers'. Consequently, around 10 per cent of the population of the town of Löcknitz in 2011 were Poles (283 of 3,070 inhabitants), similar to Löcknitz-Penkun *amt*[1] (932 of 10,737 dwellers) (Zensus 2011a) and to Uckermark (around 1,300 of 121,326 inhabitants in 2013). In the same year, in the nearby city of Schwedt 20 per cent of the doctors in the local hospital were Poles, and in the small town of Gartz 20 per cent of the children in nursery schools had Polish citizenship. On the other hand, in 2010 around 2,500 Germans were working in Szczecin, many of them commuting every day (Gazeta Wyborcza, 2010).

The presence of the Polish settlers on the German side of the border has contributed to the economic revival of the province (also creating workplaces for local Germans), infrastructural rebirth (reducing the number of empty flats in Löcknitz from 12 per cent to 1 per cent) and is also having a positive effect on birth rate, school statistics and so on. In Löcknitz a Polish-German gymnasium offers bilingual education, and the town covers the costs of a shuttle bus, transporting some of the Polish pupils to Szczecin every day. The creation of this new multicultural community is not smooth; problems result, among other issues, when some of the Polish inhabitants treat the German border towns as a suburb of Szczecin, having most of their everyday life activities on the Polish side. This results in social alienation and limited contacts with the local German population (Zanikająca Granica, 2012).

In Frankfurt (Oder) the situation is different. Local economic asymmetries have been following the 'big scale' relations. The unemployment level was about 11 per cent in Frankfurt (Oder) in 2010 (9 per cent in 2014), and 13 per cent in Słubice County. Local economic problems are alleviated by the proximity of Berlin, where many of the town's inhabitants work. The policy of the local authorities has never aimed at attracting Poles to move to the German side of the border, and real estate prices have been too high for most of them. In the face of ongoing depopulation, the local authorities preferred to demolish unattractive and empty apartment blocks located on the periphery of the town, rather than offer them at a lower price to their Polish neighbours. On the other hand, the spatial proximity and walking distances across the town border bridge do not require Poles to move to the German side of the border in order to benefit in various ways from the proximity of Frankfurt (Oder). Despite the domination of the German city within the pair, the urban structure is relatively balanced and forms a single functional organism with two components, two centres and two nations that – intensive contacts notwithstanding – are spatially segregated. This has been visible in the political concept of the 'Dual Town' (*Dwumiasto*, *Doppelstadt*), which describes its duality and exceptional nature, transcending national boundaries (Jańczak, 2011). This concept is implemented in the form of the long-lasting, common development strategy of both town halls. The only visible exceptions are the universities and the students, forming the largest cross-border community with regard to place of residence, social contacts, language relations and so on, but belonging to the category of 'temporary residents', not 'locals'. Another group of inhabitants that stresses the unity of the towns and facilitates cross-border everyday life is centred around the SłubFurt association. Spatial de-boundarization has not been followed by massive social frontierization. Consequently, only 923 Polish citizens lived in Frankfurt (Oder) in 2011 (with a total population of 59,140), forming 1.6 per cent of the city's population (where another 1,454 foreign passport holders were registered) (Zensus 2011b, p.7). In

2013 this number grew to about 1,300, still making the Poles a marginal part of the permanent dwellers.

The spatial situation of Görlitz and Zgorzelec is very similar, with the two towns within walking distance of each other. They functionally work as a single organism separated by an invisible border, which led to the concept of the European Town employed in 1998, stressing on the one hand the down-scaled European project, and on the other hand the concept of one town and two states. The economic asymmetries are somewhere in between those described in the cases of Löcknitz and Frankfurt (Oder), with visible central investments in Görlitz, but at the same time 13 per cent unemployment registered in 2010, and 9 per cent in 2014, followed by the depopulation of the town described above. Here, however, local policy has aimed at attracting Poles to move to the German side. The 'recruitment campaign' became one of the official policies of the local administration, and a wide range of apartments at price levels comparable to those in Poland was made available. One of the campaign's elements was the 'welcoming package' designed for 2012–2015, offering newcomers several privileges. It was financed jointly by the town and municipal companies as well as private businesses, and included a no-rent period for an apartment, free public transportation for the first three months of residence and several discounts related to banking services and insurance. Consequently, in 2011, 1,264 Poles lived permanently in Görlitz (with a total population of 54,441 and 608 other foreigners), forming 2.3 per cent of the town's population (Zensus 2011c, p.263).

The reaction – are there cross-border urban polities?

These various examples of cross-border urbanism, each settled within a different context of local asymmetries, reveal how spatial de-boundarization leads to different forms of border change with regard to the creation of cross-border structures. In the north, the process can be described as a market-based, 'self-perpetuating' process of cross-border metropolitan expansion; in the central part as ongoing socio-spatial isolation; and in the south as a development-based policy to attract settlers. In the latter two cases, border change is a matter of specific policies (or their lack), corresponding with the political and economic dimension of cross-border governance. What is interesting, however, are the outcomes of this process, from the point of view of the social dimension of de-boundarization and frontierization. Consequently, the question is: what is the reaction to the various models of cross-border urbanism at the German-Polish border? Are there already cross-border polities, revealing a symbolic and cognitive dimension? Due to the limited space of this contribution, testing this issue is only possible in a partial way, checking only one of the manifestations of the local reaction to cross-border urbanism, namely the local support for radical right movements in the locations. The National Democratic Party of Germany (*Nationaldemokratische Partei Deutschlands*, NPD) was selected here, illustrating a negative reaction to the ongoing processes of cross-border urbanization. Its nationalistic and anti-immigrant profile has gained a local flavour in the border area, with anti-Polish elements dominating the programme, and it has gained support mainly among the young, unemployed and frustrated electorate.

Investigating the level of support for the anti-immigrant and locally anti-Polish NPD, one can easily see the differing regularities in the regions. In the last three elections to the regional parliaments (Landtag), significant support for the party was visible in the three cases under analysis (Table 3.1). However, it was higher in Mecklenburg-Vorpommern and Saxony (where NPD members were elected to the regional parliament) than in Brandenburg. At the same time, in two of the cases the level of support was as much as two times higher in the border territorial districts of Ucker-Randow,[2] with more than 13 per cent of the votes in 2006, and Görlitz than in the respective larger regions. This demonstrates that public support for the radicals is concentrated especially in those spots where the German-Polish proximity is being tested, as in a laboratory – namely, where traditional asymmetries are (partly) reversed, and Polish settlers are visibly present or encouraged to be present on German soil.

In the case of the national elections, the support for right wing radicals is lower in all the cases (Table 3.2); however, the regularity of the border towns being much more anti-immigrant oriented in comparison to their respective regions is still visible.

To complement this quantitative data with a more qualitatively oriented analysis, some examples of electoral campaigns will be presented (concentrating especially on the 2009 and 2011 votes). The main argument of the NPD stressed

Table 3.1 NPD support at the regional and town level in assembly elections

	NPD RESULTS, %					
	*Mecklenburg-Vorpommern**	*Ucker-Randow**	*Brandenburg*	*Frankfurt (Oder)*	*Saxony*	*Görlitz*
2002*/2004	0.8	1.5			9.2 (12MPs)	
2006*/2009	7.3 (6MPs)	12.7/13.1	2.5	2.1/1.8	5.6 (8MPs)	7.5/7.3
2011*/2014	6.0 (5MPs)	12.1/12.2	2.2	2.0/2.2	4.95	9.7/9.9

Source: The author.

Table 3.2 NPD support at the regional and town level in *Bundestag* elections

	NPD RESULTS, %					
	*Mecklenburg-Vorpommern**	*Ucker-Randow**	*Brandenburg*	*Frankfurt (Oder)*	*Saxony*	*Görlitz*
2005	3.3/3.5	4.8/4.6	3.3/3.3	3.7/3.6	5.0/4.8	6.8/6.1
2009	3.4/3.3	5.0/4.6	3.4/2.6	3.7/2.8	4.1/4.0	5.3/5.5
2013	3.4/2.7	5.8/4.6	3.4/2.6	4.2/3.1	3.4/3.3	5.6/4.2

Source: The author.

various forms of threats created by the Polish settlers to the German part of the borderland. The process of the influx of Poles itself was termed an 'invasion' and was to be stopped as an unwanted phenomenon. Poles were presented as stealing German workplaces and abusing the social security system, as well as causing rising criminality levels (especially stealing cars). Poland was additionally accused of planning a new nuclear plant, creating a threat to the German border areas. The arguments seemed to be local emanations of the German-wide programme of the party, where non-European immigrants portrayed in this way (who were almost absent in the border regions) were replaced by the local 'aliens', namely Poles. The phenomenon of anti-Polonism is estimated in existing research to be widespread, with around one fifth of the German population in the border regions of Mecklenburg-Vorpommern seeing German-Polish relations as getting worse due to the eastern enlargement. This atmosphere culminated in 2011 in several acts of vandalism on cars with Polish licence plates and anti-Polish inscriptions that appeared in public spaces in Löcknitz. It is also important to note the counter campaign mobilizing both local Germans and the Polish newcomers to oppose this aggression, which stressed the positive impact of the new situation on the border. One example of the latter is the 'Perspektywa' project in Mecklenburg-Vorpommern, which distributed information about the positive impact of Polish migration (Newsweek, 2011). Polish candidates (for example, Katarzyna Werth in Mecklenburg-Vorpommern, representing the Polish-German Association Pomeraniak) also stood for local election there in 2009.

Contrary to Löcknitz and Görlitz, in Frankfurt (Oder) the visibility of the NPD was lower and the support minimal. However, some direct democracy instruments have made it possible to oppose specific developments of cross-border urbanism here as well. In a consultative referendum in 2006 the inhabitants of Frankfurt (Oder) rejected the idea of extending the existing tramline to Słubice (a project which was to be co-financed by the European Union). Here, the arguments used were not anti-Polish oriented, but opposed to the transportation policy related to the formation of a cross-border town – they were more anti-Słubice.

It can thus be stated that the sites of intensive Polish cross-border migration and presence are at the same time locations of intensive local response, where a visible share of the local German population finds it hard to accept the incomers' presence. In Löcknitz and Görlitz – with a visible Polish presence after 2007 and reversed German-Polish asymmetries – anti-Polish attitudes constitute a significant part of the political landscape, which makes social frontierization difficult and reveals weaknesses in the formation of the cross-border community. In Frankfurt (Oder), where the permanent presence of Poles is low and the asymmetries are 'traditionally' constructed, the support for the radical right is small. Here, social frontierization is based on separation, which allows a polity composed of two groups, but linked to the idea of a Double Town policy, to be constructed.

Conclusions – models of cross-border urbanism

The paper has presented and compared the border processes in three cross-border settlements in the period after the eastern enlargement of the EU, and especially after the Schengen Zone enlargement in December 2007. It has revealed spatial de-boundarization (and concomitant Polish migration to the German border towns), but, at the same time, various patterns of social (re-)frontierization. The cases of Löcknitz-Szczecin, Frankfurt (Oder)-Słubice and Görlitz-Zgorzelec represent diverse combinations of migration intensity and border asymmetries. This has resulted in three models of cross-border urbanism along this border, which can be referred to, respectively, as 'colonization', 'cohabitation' and 'reverse suburbanization'. These models can be presented in the form of a matrix, combining the two variables discussed above: the low or high level of migration from the Polish side to the German side, and the level of economic asymmetries across the border (Figure 3.1).

The model of 'colonization' is represented by Löcknitz and neighbouring Szczecin, where it is the Polish side that dominates demographically and with regard to social dynamism over the depopulated provinces on the German side of the border. The migration rate is very high, with Poles settling down en masse in the half-empty spaces of the German border counties and communities. This leads to a feeling of alienation for many of the local German inhabitants, and consequently to tensions between both communities that manifest themselves in ways that include the growing support for the radical right wing party.

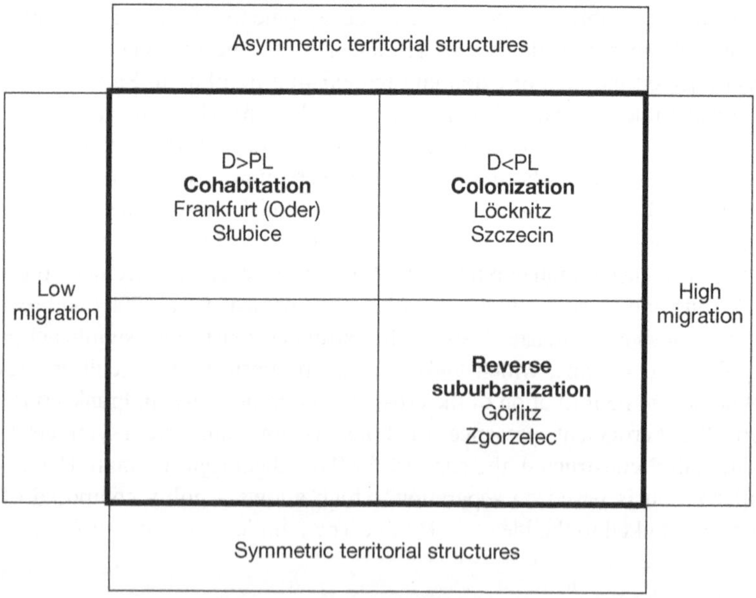

Figure 3.1 Cross-border urbanism model on the German-Polish border.
Source: The author.

Second, there is the model of 'cohabitation' represented by Frankfurt (Oder) and Słubice. 'Cohabitation' is marked by relatively visible size and wealth asymmetries. The migration level is comparatively low, due to the moderate policy on the German side. Consequently, national tensions are low, and both nations tend to reside on their 'own' side of the border and cross it intensively for commercial, professional and cultural purposes, which is the basis for the creation of a cross-border community.

Finally, there is the scheme of 'reverse suburbanization', visible in Görlitz in relation to Zgorzelec. Here, the German side dominates in terms of wealth, centrism, urban character and so on. On the other hand, it has been depopulating, which resulted in a policy to attract Poles to settle down on the German side. However, the Polish settlers have been changing the rapport between the towns, where the previously dominant and more central Görlitz is highly dependent on its 'suburb' of Zgorzelec, which reverses the relationship. The Polish presence on the German side, initiated by the local authorities, is partly opposed by the local inhabitants, which is visible in political support for the radical right.

The three tested cases thus reveal a set of regularities, where border settlements orient themselves differently to the border and border realities, as well as to the 'others' from the other side of the border. Cross-border urbanism on the German-Polish border results from spatial de-boundarization rooted in the continental integration process. However, social frontierization, manifested in the merging of previously separate communities, brings difficulties, showing that the creation of a cross-border polity is more challenging than the implementation of cross-border policies. Its intensity depends on the feeling of being endangered, resulting from the type and construction of asymmetries. Spatial de-boundarization does not automatically lead to social frontierization in down-scaled Europe.

Notes

1 The administrative division consisting of a group of municipalities, with the town Löcknitz among them. Löcknitz-Penkun directly neighbours the border with Poland, having Szczecin on its other side.

2 Former county neighbouring Szczecin, with Löcknitz in its territory.

Bibliography

Alkan, N., 2002. Borders in Europe. In N. Alkan, ed., 2002. *Borders of Europe*. Berlin: Zentrum für Europäische Integrationsforschung. pp.31–71.

Białecki, T., 1992. *Historia Szczecina*. Wrocław, Warszawa, Kraków: Olssolineum.

Browning, C. S. and Joenniemi, P., 2008. Geostrategies of the European Neighbourhood Policy. *European Journal of International Relations*, 14(3), pp.519–551.

Bürkner, H. J., 2015. Scaling and Bordering. Conceptual implications of scale-related perspectives on European borders. *EU Border Regions Working Paper*, 7, pp.1–26.

Evans, G. and Newnham, J., 1998. *The Penguin Dictionary of International Relations*. London: Penguin Books.

Frelak, J. and Łada A., 2011. *Polacy nadchodzą! ... Wreszcie! Polska migracja zarobkowa do Niemiec – analiza w przededniu otwarcia niemieckiego rynku pracy dla polskich obywateli.* Warszawa: Instytut Spraw Publicznych.

Gasparini, A., 1999–2000. European Border Towns as Laboratories of Differentiated Integration. *ISIG Quarterly of International Sociology*, 4, pp.1–5.

Gasparini, A., 2008. Do European Towns Hold the Key to Cultural Integration, Incubation? *ASA Footnotes*, 36 (9).

Gazeta Wyborcza, 2010. 'Spiegel': coraz więcej Niemców ze wschodu zarabia w Polsce. 2010. *Gazeta Wyborcza*, 21 Mar.

Germond, B., 2010. From Frontier to Boundary and Back Again: The European Union's Maritime Margins. *European Foreign Affairs Review*, 15, pp.39–55.

Gualini, E., 2003. Cross-border Governance: Inventing Regions in a Trans-national, Multi-level Polity. *DISP Working Paper*, 152.

Häkli, J., 2011. Re-Demarcating Transnational Space: The Case of Haparanda-Tornio. In J. Jańczak, ed., 2011. *De-bordering, Re-bordering and Symbols on the European Boundaries.* Berlin: Logos Verlag. pp.21–35.

Halicka, B., 2013. *Polens Wilder Westen. Erzwungene Migration und die kulturelle Aneignung des Oderraumes 1945-48.* Padeborn: Ferdinand Schöningh Verlag.

Jajeśniak-Quast, D., 2000. *Geteilte Städte und Oder und Neiße. Frankfurt (Oder)-Słubice, Guben-Gubin und Görlitz-Zgorzelec 1945–1995.* Berlin: Berlin-Verlag Spitz.

Jańczak, J., 2007. Europeanization of Trans-border Communities. The Polish-German Case. In J. Jańczak, ed., 2007. *Rediscovering Europe: Political Challenges in the 21st Century EU.* Poznań: Faculty of Political Science and Journalism, pp.77–88.

Jańczak, J., 2011. Cross-border Governance in Central European Border Twin Towns. Between De-bordering and Re-bordering. In J. Jańczak, ed., 2011. *De-bordering, Re-bordering and Symbols on the European Boundaries.* Berlin: Logos Verlag, pp.37–52.

Jańczak, J. 2013a. Revised boundaries and re-frontierization. Border twin towns in Central Europe. *Revue d'études comparatives Est-Ouest*, 44(4), pp.53–92.

Jańczak, J., 2013b. *Border Twin Towns in Europe. Cross-border Cooperation at a Local Level.* Berlin: Logos Verlag.

Joenniemi, P. and Sergunin, A., 2011. When two aspire to become one: City-twinning in Northern Europe. *Journal of Borderland Studies*, 26(2), pp.231–242.

Kaiser, R. and Nikiforova, E., 2008. The performativity of scale: The social construction of scale effects in Narva, Estonia. *Environment and Planning D: Society and Space*, 26, pp.537–562.

Knefelkamp, U. and Griesa, S., eds., 2003. *Frankfurt an der Oder 1253–2003.* Berlin: Verlag für Wissenschaft und Forschung.

Koeppen, B., 2015. The EU's internal market paradigm as de-bordering tool and dominant vision for Europe? *EUBORDERSCAPES Working Paper*, 11, pp.1–20.

Kolossov, V., 2005. Theorizing Borders. Border Studies: Changing Perspectives and Theoretical Approaches. *Geopolitics*, 10(4), pp.606–632.

Kolossov, V. and Scott, J. W., 2013. Selected Conceptual Issues in Border Studies. *EUBORDERSCAPES Working Paper*, 4, pp.1–20.

Kristof, L. K. D., 1959. The Nature of Frontiers and Boundaries. *Annals of the Association of American Geographers*, 49(3), pp.269–282.

Laponce, J. A., 1972. Experimenting: A two-person game between man and nature. In J. A. Laponce and P. Smoker, eds., 1972. *Experimentation and Simulation in Political Science.* Toronto: University of Toronto Press, pp.3–15.

Lemper, E.-H., 2009. *Görlitz. Eine historische Topographie.* Görlitz–Zittau: Verlag Guenter Oettel.

Musekamp, J., 2010. *Zwischen Stettin und Szczecin: Metamorphosen einer Stadt von 1945 bis 2005*. Wiesbaden: Harrassovitz Verlag.

Musiał-Karg, M., 2009. Słubice-Frankfurt (Oder). Cooperation in a Divided City on the Polish- German Border. In J. Jańczak, ed., 2009. *Conflict and Cooperation in Divided Cities.* Berlin: Logos Verlag, pp.150–166.

Newman, D., 2006. Borders and Bordering. Towards an Interdisciplinary Dialogue. *European Journal of Social Theory*, 9 (2), pp.171–186.

Newsweek, 2011. *Protest przeciwko antypolskim plakatom wyborczym NPD.* 2011, 17 Aug.

O'Dowd, L., 2002. The Changing Significance of European Borders. *Regional and Federal Studies*, 12(4), pp.13–36.

O'Dowd, L. and Wilson, T., 2002. Frontiers of sovereignty in the new Europe. In N. Alkan, ed., 2002. *Borders of Europe*. Berlin: Zentrum für Europäische Integrationsforschung, pp.7–30.

Popescu, G., 2012. *Bordering and Ordering the Twenty-first Century: Understanding Borders*. London: Rowman and Littlefield.

Problemy z Polakami? O resentymentach wobec Polaków w Mecklemburgii. 2009. Publikacja Regionalnej Placówki Kształcenia, Integracji i Demokracji (RAA).

van Houtum, H. and van Naerssen, T., 2002. Bordering, Ordering, and Othering, *Journal of Economic and Social Geography*, 93 (2), pp.125–136.

Waack, C. and Haslinger, P., eds., 2010. *Görlitz/Zgorzelec. Historisch-topographischer Atlas schlesischer Städte. Band 1*. Marburg-Wrocław: Herder-Institut.

Walters, W., 2004. The Frontiers of the European Union: A Geostrategic Perspective. *Geopolitics*, 9(3), pp.674–698.

Wesołowska, E., 2012. Coraz więcej Polaków we wschodnich Niemczech. Dostają wyższe pensje, tańsze domy i przywileje socjalne. *Gazeta Prawna*, 4 May.

Zensus 2011a. Bevölkerung nach Staat der Staatsangehörigkeit und Geschlecht für Gemeinden Ergebnisse des Zensus am 9. Mai 2011, Statistische Ämter des Bundes und der Länder, Erschienen am 10. April 2014, Statistisches Bundesamt, Wiesbaden 2014.

Zensus 2011b. Bevölkerung und Haushalte Kreisfreie Stadt Frankfurt (Oder), Stadt am 9. Mai 2011.

Zensus 2011c. Bevölkerung im Freistaat Sachsen am 9. Mai 2011 nach demografischen Grundmerkmalen, Endgültige Ergebnisse Gebietsstand: 1. Januar 2014, 2. Korrekturausgabe, A I /Zensus 2011–3.

Znikająca Granica, 2012. Warszawa: Instytut Spraw Publicznych.

Part II

Communities, relationships and identities in borderlands

4 What makes a place

Traces of the border in rural villages affected by cross-border residential migration

Gregor Schnuer

Framing cross-border dwellers

The Maastricht Treaty, which introduced the so-called 'Citizenship of the European Union' and was signed in 1992, together with the Schengen Agreements (1985 and 1990) which removed national border controls between signatory states, ensured European residents the right of free movement, settlement and employment in member states of the European Union. In spite of this freedom, it has been noted that persons, especially compared to goods and services, are relatively less mobile (van Houtum and van der Velde, 2004; van Houtum and van Naerssen, 2002). The borders, which continue to exist as national, political, social, cultural and economic demarcations, although admittedly to varying degrees, appear to continue to inhibit the movement of persons in a very real way. Even though the observation of limited movement is generally true, it does not apply as clearly to borderlands as it does to observations made on the national scale. Borderlands in Western Europe have shown relatively high levels of mobility historically, something that has increased after the Schengen Agreements and Maastricht Treaty. They offer us the sites where the European Union's political push towards integration most explicitly affects the everyday life of residents. However, borderlands differ greatly from one another, and what exactly the term 'borderland' describes remains contentious (Häkli and Kaplan, 2002).

The particular role of the relative proximity of residents on either side of the border to those on the 'other side' has been frequently noted in border studies. As such, borderlands have been described as 'laboratories' (van Houtum, 2000, p.65) and even as complete conceptual alternatives to 'borders', since the notion of borderlands moves away from a fixed line, instead emphasising ambiguity (Hassner, 2002). It is exactly in these spaces, which flank national demarcations, that identities can be both dichotomised by heightened self/other distinctions and confrontations and problematized through shifting national boundaries, cross-border regional identities and more frequent mobility. Borderlands are, therefore, deeply heterogeneous and the economic and social relationships, the everyday practices and the perceptions and experiences that are shaped by the proximity to the border (Balibar, 2009; Rumford, 2006) play out in different ways depending on the context of the particular border and the specific border region. The following chapter takes both this particularity of borderlands and the ambiguity of the role of the border within them as points of departure to explore how they manifest in the way in which the autochthonous population of villages in close vicinity of the border perceive their place, their home, to be changing, and how this change is related to borders and boundaries.

In human geography and sociology, frequently if not overwhelmingly, borderlands are researched as sites of cross-border mobility. In many ways the often high frequency of movement, whether higher or lower than expected, back and forth across the border, is understood to be central to, and even a defining characteristic of, borderlands, especially European borderlands following the Schengen Agreements. These movements are a result of, and also result in, the blurring and complicating of national borders. For example, if we take a short glance at research on the Dutch-German borderland near the city of Nijmegen, we see that mobility (and its absence) is the key focus of research questions. Consumer mobility (Spierings and van der Velde, 2008; Spierings and van der Velde, 2013), labour mobility (van Houtum and van der Velde, 2004; Knotter, 2014) and residential mobility (Strüver, 2005a; Strüver, 2005b; Gielis and van Houtum, 2012; van Houtum and Gielis, 2006) are researched as phenomena that offer insight into the manner in which the borderland ambiguously both exhibits and defies a clearly defined national border by reference to mobility or movement across the border.

It is particularly the latter form of mobility in the above list, namely cross-border residential mobility, that has received growing attention, in part because it is a recent form of mobility that has begun to occur at an increasing number of Inner-European borders over the last decade (Jagodic, 2012). Research on this form of mobility, which sees residents from one side of a border relocate their home, often only a short distance, across the national demarcation to live 'abroad', has yielded several innovative conceptual ways of understanding the border in the context of borderlands. Given that persons that relocate a short distance across a national border technically become migrants, but at the same time in almost all cases maintain a workplace, friendships, family relationships and other important everyday ties and thus in many ways do not leave their place of origin, the usual conceptual repertoires of migration studies and globalization studies fail to capture the ambiguity and complexity of the particular form of mobility of these 'migrants'.

Conceptually capturing these people's movements across the border often focuses on playfully engaging with questions about the nature of mobility and migration, the dichotomy of inside and outside, and the social construction of space and place. One such attempt to bridge conceptual gaps is the use of the term 'elastic migration' to describe those that migrate across a border but stay within a borderland (van Houtum and Gielis, 2006). Van Houtum and Gielis identify this elasticity in the close connection to the country of origin that, unlike transmigration in the typical sense of long distance migration, takes the form of frequent, almost daily, border crossing. This continuous crossing questions the presumed equivalence of nation and state, or so they argue, since 'there is a difference between respecting/adopting the formal rules and laws created by the … state or municipal authority and respecting/adopting the customs, rituals and conventions of the [receiving] people in the border places' (2006, p.196). This distinction between formal rules and lived customs and conventions can be related to a more fundamental question that is raised by the phenomenon of cross-border residential mobility, namely, 'what is it, to dwell in a place?' Here, this includes a concern for how a place is constituted by the people who dwell there and their practices.

More recently, Gielis and van Houtum (2012) also tackled this question, at least from the perspective of cross-border residential migrants. They attempt to come to terms with the ontological and epistemological quandaries of this 'choreography of the transborder dwelling' (2012, p.799) by trying to bridge two juxtaposed understandings of dwelling: the understanding of dwelling as rooting and b/ordering on the one hand, and the more nomadic understanding of dwelling as routing and de-bordering on the other. They bridge this gap by drawing on the work of Peter Sloterdijk, reimagining the house, as a dwelling (across the borders), to be both a 'bubble' and 'foam'. The house is a sphere that serves as ordered shelter, yet at the same time this bubble is becoming increasingly internally textured through connections to the 'outside' – that is, by bringing outside inside, for example through the use of telecommunications technologies. Houses, according to Gielis and van Houtum's reading of Sloterdijk, are 'multi-relational and multi-local spheres which are constantly changing in form and content' (2012, p.810). This idea of a foamed sphere blurs the boundary between an inside and outside – in this particular case, it is the inside and outside of the house, but it is also a conceptual understanding of a place that can be applied to other inside/outside relationships.

Gielis and van Houtum argue that the foamed nature of the house as dwelling can also be applied to a borderland setting. Even though this transposition of the concepts of bubble and foam from house to borderland in their work is not entirely convincing and arguably underdeveloped, one corollary from this blurring of boundaries, especially in relation to ideas about dwelling, is the implications for ideas of mobility, given that, conceptually, mobility tends to describe movement that crosses some form of boundary. Gielis and van Houtum's connection of monadic and nomadic understandings of dwelling touches on the issue of the boundedness of a place, something that is, as I have argued elsewhere (Schnuer, 2014), intricately connected with movement and the distinction between those movements that 'leave' a place and those that 'stay' somewhere.

The following chapter continues this challenge to typical understandings of places as bounded containers by exploring dwelling, mobility and place in the context of the phenomena of cross-border residential mobility. Unlike previous work on the topic, which either focuses exclusively on the 'migrants' or on the relationship between the new and old residents, the following paper will focus on the autochthonous population of villages that have been clearly affected (demographically) by the influx of border-crossing residents. Through a detailed analysis of two interviews, the chapter will present two answers to the question 'What defines a village?' These answers will allow us to consider, on the one hand, the borders between the home and the village, the village and its outside, and the home country and the neighbouring country and, on the other hand, how, if at all, residential migrants are perceived to play a role in the blurring of any of these borders. Overall, this contribution hopes to show how the crossing of a border, such as the case of residential migrants moving into the village, also affects other, more local and immediate boundaries in interesting ways.

The Borderland in question

Luxembourg and its surrounding borderlands exhibit several particular characteristics in comparison with other European borderlands, including other areas where we can see a growth in cross-border residential mobility. Most of these particularities are the consequence of two features of this border region. First, Luxembourg is an economic hub and attracts large numbers of cross-border commuters and international migrant labour. Second, this hub is not situated at a dichotomous national border – like, for example, Szczecin in Poland (another example of a regional economic hub situated along a single national border) – but instead, due to the small size of Luxembourg, three national borders are in very close proximity; each bordering country – Belgium, France and Germany – is no further than 20km from the capital city.

With regard to the economic role of Luxembourg in relation to the cross-border labour market, the most telling statistic is the overall number of cross-border workers, which in 2014 amounted to 160,000 (STATEC, 2015a), which represents approximately a quarter of all cross-border workers in the European Union (OIE, 2014). This means that only a small majority (around 53 per cent) of people employed in Luxembourg also reside in Luxembourg. Additionally, the small country is also internally demographically diverse as it attracts large numbers of employees from abroad, with non-nationals making up 45 per cent of the population in 2014 (STATEC, 2015b).

The fast influx of foreign residents into a small country increases the demand for housing and this, taken together with the relative affluence and very high average salaries, has led to a steady and steep increase in real estate prices in Luxembourg. Even though these prices have increased notably in the bordering regions over the last decade, the explosion of real estate prices in Luxembourg has been relatively well contained within the national borders. As a result, the surrounding areas offer often dramatically cheaper housing, and this difference between Luxembourg and its neighbouring regions, much as with other examples of cross-border residential mobility, has given rise over the past fifteen years to increasing numbers of people relocating from expensive Luxembourg to its affordable surroundings.

The composition of these residential migrants, given that Luxembourg is demographically diverse, is also more complex than in other examples across Europe. Even though precise data on the number of residential migrants and their composition in terms of nationality, age and employment do not exist, the available sources suggest that we can roughly identify four groups (which overlap to some extent): Luxembourg nationals, who predominantly choose the German borderland as their destination; residents who are nationals of the surrounding countries, either from the borderlands or from further away, who mostly moved to Luxembourg for work and who now return to their home country; residents who are typical labour migrants (including second and third generation), who predominantly move to francophone regions (Wallonie in Belgium and Lorraine in France); and lastly residents who are part of the 'international elite' who work in Luxembourg's financial and business sectors, who predominantly relocate to the German borderland, like Luxembourg citizens.

The research presented here was conducted in four villages on the German side of the common border with Luxembourg along the Moselle River. Our empirical research consists of open-ended interviews and participant observation, talking to both residential migrants and autochthons, even though this distinction itself is difficult and varied, as shown, for example, by the fact that some interviewees who moved from a neighbouring German village some twenty years ago do not consider themselves fully locals.

As a reaction to the continuously growing demand for properties in the region, many villages have begun to provide land for sale by transforming adjacent farmland into developed plots. All the villages that we looked at in our research have witnessed a considerable population growth due to the influx from across the border. As an example, over only a few years one village turned from a distinctly rural locality with an almost exclusively German – and autochthonous – population into a cosmopolitan community with more than forty nationalities. Even in small villages where we did not find the same diversity of international residents, we nonetheless found proportionately very large residential developments, some doubling the number of residential houses. In addition to the most visible change – new residential developments, where several plots are made available at the edge of the village and mostly new residents build detached houses – there is also another change: a smaller number of new residents have purchased old houses in the heart of the villages, some of which would otherwise be derelict, often restoring them.

Lastly, the autochthons frequently mention that, despite all these changes in the village, especially the larger number of inhabitants, little has changed with regard to social and communal participation, where they note that it is the same familiar faces who attend local events. There is a juxtaposition of the growing population and the dwindling or stagnating participants in social activities – we will return to this seeming contradiction and the resultant disappointment of some autochthonous expectations.

Saardorf,[1] the village which is home the two men whose interviews are presented here, is a small village with around 100 inhabitants in 2011. Around this time a new residential development was begun, effectively doubling the number of houses. This was not the first residential development: in the 1990s a small neighbourhood of just over a dozen houses was developed. Saardorf is not only small, but it also lacks amenities that larger villages have to offer: there are no shops, no baker, no butcher, no school or kindergarten, no pub and no café. The village's everyday social interactions are not focused on public spaces where residents will cross paths frequently. The communal aspects of the village, as will become clearer through the examination of the interviews, is instead based on 'neighbourliness' and a few social events that happen throughout the year.

To summarize, Saardorf, like most of these communities, was stagnating or shrinking in size, as well as aging. As in many small rural villages, increasing numbers of old houses were vacant and falling into disrepair and the last remaining amenities – small grocery stores and village pubs – were disappearing. Harald, a resident of Saardorf, uses the word *lebensfähig* – being viable – to describe this threat of dying out, a threat that could be avoided through the high demand for building plots from across the border, allowing the village to attract new residents easily. On the other hand, even though the influx of many new residents and the rapid expansion of these villages is technically keeping these places 'on the map', there remains a different threat – not to the viability of the village, but to the social and communal structures that give the village its particularity and differentiate it from a suburban residential area. Matthias, another resident of Saardorf, is explicitly worried about these structures, their current decay and how they are affected by the recent changes and overall constraints of small rural villages. The interviews with Harald and Matthias will be explored in detail to better understand how diverging understandings of a place relate to attitudes about the changes that the village of Saardorf is experiencing in its situation in a fast changing borderland.

The flourishing place

Harald, the resident concerned with Saardorf's ability to survive, offers a detailed perspective on the new residential development and the village as a place. Now in his sixties, Harald is active in local politics but was not born in Saardorf; he moved there forty years ago. Formerly a farmer, he has since changed his profession. At first he continued to farm as a secondary occupation while working in the agricultural branch of a regional employer. Eventually he and his wife also developed and gradually diversified a local business, adapting to changing socio-economic conditions and opportunities. Their venture is talked about at length, especially the renovations and alterations that the farm's change in use required, which are described as part of a proactive and practical approach to dealing with the transformation of rural livelihoods. It is quite clear that he understands himself as a person who is willing to accept the need to adjust, especially given changing circumstances, such as the declining viability of farming as a full-time profession. However, this attitude is not just biographically relevant, since it can also be seen in his approach to his political engagement.

Harald, as one of the decision makers behind the new residential development, speaks of it as a great opportunity for a small village like Saardorf, which in his estimation was slowly fading away. In many ways, this concern for something in Saardorf being lost or diminishing is not too different from Matthias's worries, but we will return to the comparison between the two residents below, once we have considered both of them in detail. For now, let us remember that Matthias was critical of the new residential development, a sentiment that Harald is keenly aware of, noting that this decision caused divisions among locals. Nonetheless, unpopular and tough choices are necessary to see Saardorf become a flourishing village. After all, his approach to reviving the village, which seems to be exemplified in the new residential development, results in opposition because there are those residents who have a parochial 'village mentality' – they are inflexible and struggle to adjust to the changes that are, to him, inevitable. Autochthons, in Harald's interview, appear mostly as obstacles to reviving Saardorf, since they do not understand the need to adapt. To paraphrase Harald, the village was no longer *lebensfähig*, and it was imperative to the place's viability to attract new residents, not only for the sake of numbers, which matter administratively to some extent, but also, and more importantly, for the continuation of the community that constituted the place.

The way in which Harald frames the necessity of change is indicative of how his perspective places the village within a broader changing context: rather than transforming the village, Harald seems to envisage adapting it to broader changes that are beyond its control. This includes the dwindling farming community – there remains only one farmer in Saardorf – and the tendency of young people to move away. He draws parallels between his own life and the village with regard to this: being a farmer was no longer viable as a profession, so he had to begin something new, which he describes as a difficult transition. We can reassess Harald's autobiographical exposé and consider it, in part, to be reactive – that is, a flexible way of dealing with circumstances beyond his control. Put differently, in his own estimation he has been able to make the best of a difficult situation, both personally and with regard to the village.

Harald talks extensively about the new residents and the many adjustments that may be seen as corollaries of the new residential development and the influx of many strangers from across the border. The way that new residents feature allows us to see how far-reaching Harald's understanding of adaptation goes. There are a plethora of individual interactions between himself and the new residents as a result of his active participation in village politics. These make clear that he is not only interested in accumulating new residents, but that he is also attempting to accommodate them – that is, to encourage them to feel as though they are a part of the place. Concretely, this involves inviting a new resident to his home for a glass of sparkling wine in order to appease a conflict between children who played an innocent prank, or bringing new residents a small gift on behalf of the municipality to congratulate them on the birth of their child. These gifts are important because of the need to make new residents feel welcome and to give them a positive impression of the village. This motivation behind such gifts is also the reason why Harald considers them less necessary for autochthons.

Welcoming new residents does not only involve personal contact; Harald also speaks of some changes to the traditions. These include offering a wider variety of food and drink at events, and even moving an event that was traditionally held on a specific date to the nearest weekend in order to allow young working parents to attend more easily. Again, these changes are attempts to appeal to the new residents, and it is worth noting that these ways of facilitating participation in the community are not the same as attempts to offer potential new residents something appealing to attract them to move to Saardorf rather than another village (which other village representatives spoke about with regard to infrastructure). Harald's forthcoming efforts are directed towards people who are already residents, but who, to him, are not quite part of the village yet. We can juxtapose a phrase frequently used by autochthons to speak about residents that do not participate in village life – 'they only dwell here' – with the ideal resident that Harald's efforts are attempting to encourage – efforts which he refers to as an obligation.

It would be premature, however, to see in this eagerness to accommodate new residents a simple division between old and new residents in general. Instead, the interview reveals a complicated mixture of distinctions on the one hand and a frequent simplification consisting of equating new residents with Luxembourgers on the other. Harald's efforts are not only an effort to make new residents feel welcome; they are in part also attempts to make specifically Luxembourgers feel accommodated, for example by offering sparkling wine at a local event, since the former is stereotypically popular with Luxembourgers. The manner in which these changes are discussed makes it clear that Harald considers 'the Luxembourgers' to have different and in some sense more demanding tastes, and in order to make local events attractive to them there must be something more on offer than would usually be the case. Even though these changes appear innocent, they do show clearly that Harald has a particular and quite concrete idea of who the new residents from Luxembourg are, or at least, what they are like. Harald speaks about the satisfaction and contentment of these Luxembourgish residents, considering them both possible new members of the community and residents who serve the role of ensuring the viability of the village.

The nourishing place

We will continue by looking at the concerns of Matthias, who is similar to Harald in many ways: he is in his early sixties, active in both local politics and the village societies and clubs, and also comes from a farming background. Unlike Harald, Matthias was born in Saardorf and still lives in his family's renovated farmhouse. Throughout the conversation about the village, its current situation and future, the general tone can best be described as concerned. In particular, he is concerned about preserving the village. He discusses three ways in which Saardorf is at risk of losing something vital to its identity and particularity. In other words, Matthias believes Saardorf is at risk of becoming a different, less desirable version of itself.

The first example of this concern for the village relates to the *Dorfbild*, meaning the village's overall appearance. For Matthias, this includes tidy front yards and swept pavements on the one hand, but on the other it also concerns architecture. The maintenance of the appearance of the village is considered a communal responsibility, especially since the refusal of some to participate is enough to spoil the whole look – if, say, two or three houses have overgrown and ill-kept front yards. He expresses great respect for the new residents who, rather than buying a plot to construct a new house, purchase an old vacant home in order to restore and renovate it. In Saardorf, as in other small villages, the old homes sold to new residents vary in size and style and include large farmhouses and even the old rectory. Some of the new residents are beginning to play a positive role in this context, since without their commitment – and he notes that renovating an old farmhouse requires money and effort – there might come 'a second generation here after us, who will no longer know what an old farmhouse looked like, and there will only be one-storey, well insulated houses standing around ... but everything else that was significant around here for two hundred years before will be gone' (25:51). Matthias is not concerned with the viability of the village – that is, its number of residents or how many people attend local events. Instead, he is concerned with a very particular version of Saardorf. He does not seem to question whether there will be people living in houses in Saardorf in the future, something that is not in itself significant. Purchasing an old house in the village, rather than a new construction in the new residential development, is seen as participation in something that is significant and valuable. These residents take on a constitutive role in Matthias's understanding of Saardorf as a particular place, and they are contrasted, at least implicitly through the juxtaposition of newly-built bungalows and traditional farmhouses, with the new residents whose impact on the *Dorfbild* should be considered undesirable. Dwelling in Saardorf, for Matthias, evidently expands beyond the constraints of the house. We might even go so far as to claim that for Matthias, 'to dwell' includes a commitment to and participation in the particularity of the village of Saardorf.

Second, Matthias is concerned about the autochthons, especially the old and the young. The elderly, including his own mother, rely on communal structures such as meetings in the graveyard or the church choir. The elderly meeting at the graveyard, tending to the graves of spouses and relatives, is particularly noteworthy because it shows a general rootedness in the place. The meetings of elderly people are not just described as an occasion to catch up with one another and socialize; there is also a subtler relatedness that can be read into Matthias's account. He mentions not only that his mother meets old friends, but also that she knows who is at the graveyard at about what time. Furthermore, unlike new residents, residents born in the village experience the graveyard with genealogical familiarity and connectedness. More generally, Matthias recounts the meetings at the graveyard in such a way that this is an example of something that is important to the elderly, something that makes Saardorf their home.

With regard to the church choir, he explains that even if the number of members was dwindling, it would not be an option to merge the small choir of Saardorf with those of neighbouring villages, because the autochthons who have spent thirty or even forty years participating and constituting the choir would drop out since their attachment is not just to the activity of singing, but to the communal and social aspect of the bi-weekly gatherings in Saardorf. 'It would, for me, be a sin against those people ... it would tear down structures that one doesn't want to tear down,' Matthias says (45:25). We again encounter a version of rootedness and the importance of structures that autochthons have constituted and been committed to for decades. These two examples clarify how Matthias's version of Saardorf regards the conservation of the past as integral to Saardorf's particularity, in part because the autochthons, especially in the case of the graveyard meetings and the church choir, are both constitutive of and constituted by that past.

Even if the increase in residents might lead to a point at which new infrastructure could become economically feasible, such as a small café or a shop, this would not replace or substitute for these old structures. Even though there is a café or similar space to meet and socialize and a choir in a neighbouring village a few kilometres away, providing a possibility to sing for new residents, I would argue that Matthias's concern for the structures that the autochthonous elderly rely on could be described as a concern for their ability to 'dwell' in Saardorf, provided we understand dwelling, in part at least, as being constitutive to a place and involving a broader participation in the community than just a house to live in.

It is not only the older generation whose 'dwelling' is threatened, however. The younger generation, for Matthias, is experiencing this threat much more literally, since they are unable to stay in the village due to a lack of available houses and building land. Matthias mentions that the new development regrettably only reserved three plots for 'our own people' (05:32). While there is no current demand for plots of land from autochthons, Matthias mentions that only a few years ago, when there was no available building land, some young residents who wanted to stay had to relocate to another village. Here we find a more explicit connection between new residents from Luxembourg, for whom, in the eyes of some autochthons including Matthias, the new residential development in the village was meant, and concern for the village. Matthias expresses a sense of responsibility for the younger generation that has grown up in Saardorf, and he frames their departure as a loss for the place. It is even possible to connect the way in which the old residents are shown to be deeply rooted in the village and the way in which excluding the old and the young, by displacing them either to their homes by destroying their communal structures or to neighbouring villages by failing to provide homes, threatens the identity and particularity of Saardorf for Matthias.

The third and final example of Matthias's concern for the village shows us yet another way in which dwelling can be understood as constitutive to the sense of particularity of a place. He explains how in order to get things done in a small rural village, residents do them themselves, through their own personal commitment and effort. There are two sides to this. First, given the small number of residents and the rural location, there is little funding and support for public improvements from the regional administration, so these works are done using voluntary labour with material costs funded through the village administration or donations. One example of this is the construction of a small playground by the members of the village society, a voluntary club for the improvement of Saardorf. Further examples include the organisation of the few items on the village's social calendar, such as the small gathering for the raising of the maypole, a regional tradition, and a larger fair that is particular to Saardorf and attracts visitors from outside the village. The former event preceded the interview by a few weeks, and this year, for the first time, quite a number of new residents joined in, eating grilled meats and drinking a beer or two. Matthias mentions two new residents in particular who asked to help. After directing them to some work in the kitchen, he told them 'we need this [the active helping] here with us, so that we don't just bring the full bottles and you drink them dry, but that *we* also carry them back'. Second, this 'do-it-ourselves' approach is not limited to public events that can be considered communal, but also includes the mutual and reciprocal exchange of machinery, tools, experience and assistance among residents when working on their own homes. This aspect of dwelling in these villages is frequently mentioned and also juxtaposed to paying to get the work done, which is a solution associated with new residents, particularly those who work in Luxembourg and have salaries high enough to afford professional services.

We can summarise by stating that Matthias seems to be concerned about the village's aging autochthonous families, their homes, their day-to-day interactions and connections and the appearance of the village as a whole. More concretely, this concern focuses on preserving the past by ensuring continuity in these aspects. Bearing this in mind, we might argue that it is not surprising that he does not speak about the new residents in great detail. They almost exclusively feature when they are participants in this continuity – that is, when renovating houses or lending a hand at a local event. The new residential development and the accompanying influx of new residents threaten continuity in Matthias's view, but not only because there are many strangers now and many homes that are a bad fit stylistically. Matthias also thinks, or so I would argue, that the new development threatens continuity by distracting the local political focus away from what matters and towards the participation in a short housing boom where only some locals might stand to profit from selling land. The decision for the new development was made behind closed doors, he says, and this exclusion of the community and the lack of attention paid to the way in which this large expansion might affect the village might be as significant a concern for continuity as the new development that resulted from it. Matthias evidently struggles to see any advantages of the new development, and even potential benefits, such as an influx of revenue for the community, have, according to him, not been realised, as the plots were sold at too low a cost. In this view, therefore, Saardorf is beginning to transform into a different place, in which the residents no longer form a nourishing community and where the village is no longer a dwelling place that nourishes a rooted identity which is intrinsically connected to the village, its residents, its appearance and its practices. In an almost resigned manner, Matthias concludes that 'time will tell' with regard to the effects and consequences of the changes brought about by the many new residents, and that both sides, old and new residents, will have to learn to accommodate one another.

Saardorf – a tale of two villages?

If we compare Matthias's perception of Saardorf and the village's changes in recent years with Harald's account of his efforts to change the village, we might, at first sight, consider these two residents to hold antithetical views. However, this conclusion would be premature. I want to begin this comparison by outlining a key similarity that might easily be overlooked. Both Harald and Matthias consistently speak about living in a rural village in a very similar manner; their observations are not about living in a house that is considered a 'home', but about living in the village. In fact, residents who only dwell in their houses are often considered to barely be residents at all. This was made particularly evident through the repeatedly clarified importance of participation and interaction in the community. In Matthias's depiction we can see a concern for the capacity and ability of the older and younger generations to participate in their community, whether by membership of a church choir that brings the elderly together or by ensuring that younger people are able to afford a home. In much the same way, although orienting exclusively towards new members of the community, Harald actively tries to facilitate participation in the community. He hopes to get new residents out of their houses and into the village. In both men's concerns, and in the ways that they speak of dwelling in the village through practices, we can also see that they share much of their understanding of what it means to dwell in a rural village: dwelling in Saardorf is something that happens not only in the private home, but also through the interactions between residents and the connections between the home and village that these interactions result in, be it an official gift for the birth of a baby or public interest in the state of individual front gardens.

In some of these accounts we also begin to notice something different – namely, that the boundary between the house and the village as dwelling place is, as described in the beginning, not a barrier. Instead we see a membranous connectedness between the home and the village, where, for example, Harald drops off a gift for a newborn child, or neighbours comment on the condition of the home and its garden. The autochthons can also be said to expect new residents to dwell in a foam-like manner, where the home is connected in ways that permits the outside to enter and the inside to exit. Unlike the case described by Gielis and van Houtum (2012), Harald and Matthias recognize the foam-like reality of dwelling in their immediate surroundings, rather than in cross-border interactions. There are no telecommunications at play, but a short walk to the neighbours. Inversely, it is also true that Harald and Matthias are not preoccupied with the new residents' connections across the border (e.g. family, friends, etc.).

In spite of their differences, Matthias and Harald nonetheless resemble each other in their understanding of what it means to dwell in a rural village. Their participatory notion of dwelling is, so to speak, a traditional version of the aforementioned understanding of the home as a bubble and a foam. Furthermore, the participatory dwelling at the heart of this foam-like structuredness may be extended from the house to a different level of dwelling place, the village, which appears itself foam-like, internally structured, yet membranous in its relatedness to the region, neighbouring villages and so forth. It could even be argued that Matthias and Harald offer some insight into a plausible transposition of the concepts of bubble and foam to a broader understanding of dwelling place, albeit focused on sedentariness.

Given their shared understanding of what dwelling involves in this context, it might be less obvious why Matthias and Harald have very different attitudes towards the new residents and how the village is changing. For Harald, the new residents from across the border have a much more important role to play in the future of the village than for Matthias, for whom they are an ambiguous addition to the town, neither part of the solution to the problems that he identifies, nor the causes of these problems. Even though a further similarity is that both situate the village in the wider region that is changing – small villages are shrinking, old farmhouses are standing empty and farming is shifting – I would argue that one source of their diverging perceptions is that they differ fundamentally on the relationship of Saardorf with this region.

Matthias, who speaks of the typical regional farmhouses as things that are important to maintain, seems to consider Saardorf important in part because of its distinctly rural and regional character. For him, maintaining this distinct form of community requires a concern for the autochthons and the capacity to rely, to a considerable degree, on one's own resources and strengths – problems such as infrastructure regeneration have to be solved internally, by residents putting in the effort. The rapid regional change, which in recent years has brought about the arrival of new residents from across the border, does not offer a viable solution for Matthias because he sees it as accelerating the degradation of the distinctly rural character of Saardorf. He does seem to agree that some new residents are important, but he believes the influx should happen slowly and cautiously. This may also be connected to the way in which the situations he cites rely on maintaining and fostering a sense of rootedness in the village, something that plays a role in both his account of the young people who are unable to stay where they were born and old people who are unable to continue decades-old practices. This focus on Saardorf in particular, and the depiction of rural villages as somewhat isolated in several ways, suggests that even though Matthias regards dwelling in a house in Saardorf as similarly foam-like to Harald, his understanding of the village in the context of the region is more clearly bounded and bubble-like, making the blurring of the boundary between Saardorf and its surroundings, especially Luxembourg, part of his concerns.

Harald's perspective of the region is different. He is less worried about a way of life that needs to be saved, but he is concerned with keeping the village flourishing at a time when its distinctly rural way of life is, in part, what is causing the problem. Whereas Matthias sees Saardorf as intrinsically connected with the regional and rural way of life, Harald seems to consider them separate; much of the way of life is disappearing and change is required to make sure that the village does not disappear along with it. The fact that these border villages are able to attract new residents from across the border with relative ease due to their particular border situation is not a threat to the village, but a positive opportunity to turn shrinking communities into renewing villages. He recognizes that selling plots of land to Luxembourgers does not result in a flourishing community over night, but he seems convinced that, with enough effort, it is possible to achieve this as a long-term goal. However, the changes that Harald is not only willing to concede to but actively pursues might change Saardorf in a way that Matthias considers part of the decay of the community. By attracting residents from across the border, Saardorf is thriving because of the new residents who nevertheless live cross-border lives only insofar as the autochthons can make the village a place for new residents to dwell in a participatory manner. If they do not achieve this, their new residents will 'only sleep here'. Harald seems to desire Saardorf to open up, to be arguably more foam-like in its connectedness to opportunities that lie beyond the village itself.

Because their disagreement can be traced to a different understanding of the village as either more bounded or more foam-like, it is relevant to the discussion of cross-border residential mobility and the particular circumstances surrounding Saardorf's location in the borderland close to Luxembourg. The dynamics in the village that result from the border, both with regard to opportunities of autochthons to earn high salaries in Luxembourg and the ability to attract new residents with relative ease, are particular to the location of the village close to this border, and afford villages like Saardorf concrete results of 'opening up' or pursuing Harald's foam-like vision. Even though neither Matthias nor Harald speak extensively about the border, or about Luxembourg and Germany as entities that constitute this border, their differing concerns about Saardorf as a rural community nonetheless show that the border is relevant to their perceptions and aspirations for their village. More specifically, the high demand for building land is most acutely relevant in their discussion of the village, and it is also one of the subjects about which the two disagree.

The men's different approaches to maintaining the village, which both rest on similar versions of the home as place but different versions of the village as place, also connect with their different perceptions of the role of new residents. Significantly, this difference is not explained by reference to distinctions drawn along national lines, such as a particular Luxembourgish mentality or way of being. These residents from across the border are therefore not regarded as blurring a national boundary or risking a national identity. Instead, we see how Harald regards the new residents as people who will, in the near future, dwell in Saardorf in the participatory way that was outlined above, and through the more membranous boundary of this rural community the new residents will make their home and come to dwell in the village. On the other hand, Matthias seems to regard this participatory form of dwelling as something that is closely connected with a rootedness, shown in his emphasis on examples of autochthons who have lived in the village for decades, and in his concern for 'long-established structures' that are not easily replaced. For him, rootedness seems to relate to boundedness, and therefore he doubts that the rapid expansion and the efforts to accommodate new residents will result in this rootedness. While both see the particularity of Saardorf as a place as being connected to the way in which residents partake in communal activities and dwell in the village, they diverge in their assessment of the village's current expansion because they seem to estimate the new residents' ability to dwell in Saardorf very differently. To phrase this another way, Matthias and Harald are most concerned with the border or boundary between the private home and the village, and they both desire the new residents to adopt the foam-like blurring of the boundary between the home and village. It is the boundary between the village and the region, especially across the border, that informs their differing views on the new residents and the future of Saardorf. The blurring of the national border, via residential mobility, impacts the boundaries across other levels or scales, such as the village and the house. This might not always be apparent at first glance because, as is the case here, the depictions of these impacts lack the semantics of national identities and border dichotomies.

What has hopefully been shown is that residential mobility in these communities is not only a concern with changing real estate prices or proportionately large numbers of new residents appearing. These new residents also bring differences in the perceptions of a place among autochthons to the foreground. They create new unusual circumstances, resulting in conflicting opinions about both the scale and speed of expansion, and fundamental questions about the purpose of the expansion, including normative concerns about responsibilities towards a place and its residents, old and new. It is interesting that the border situation fails to take on the guise of a dichotomous national divide between the old German and the new Luxembourgish residents, where the latter threaten some form of a national integrity or identity. Instead, we see the figure of 'the Luxembourger' appear in relation and juxtaposition to something much more local: the particular identity of a small rural village. Even this has been quite clearly shown to be heterogeneous and anything but straightforward, as 'the Luxembourger' appears both as a potential participant in the village's particularity and a contributor in its disappearance.

Note

1 The real name of the village has been replaced with this fictitious name in order to ensure the anonymity of participants, whose names have also been changed.

Bibliography

Balibar, E., 2009. Europe as Borderland. *Environment and Planning D: Society and Space*, 27 (2), pp.190–215.

Gielis, R. and van Houtum, H., 2012. Sloterdijk in the House! Dwelling in the Borderscape of Germany and The Netherlands. *Geopolitics*, 17(4), pp.797–817.

Häkli, D. and Kaplan, J., eds., 2002. *Boundaries and Place: European Borderlands in Geographical Context.* Lanham: Rowman and Littlefield Publishers.

Hassner, P., 2002. Fixed borders or moving borderlands? A new type of border for a new type of entity. In J. Zielonka, ed., 2002. *Europe Unbound: Enlarging and Reshaping the Boundaries of the European Union*. London: Routledge, pp. 38–50.

Jagodic, D., 2012. Living (beyond) the Border: European Integration Processes and Cross-border Residential Mobility in the Italian-Slovenian Border Area. In D. Jagetic Andersen, M. Klatt and M. Sandberg, eds., 2012. *The Border Multiple: The practicing of Borders between Public Policy and Everyday Life in a Re-Scaling Europe*. Furnham: Ashgate, pp.201–218.

Knotter, A., 2014. Changing Border Regimes, Mining, and Cross-border Labor in the Dutch-Belgian-German Borderlands, 1900–1973. *Journal of Borderlands Studies*, 29(3), pp.375–384.

Observatoire Interrégional du Marché de l'emploi (OIE), 2014. *Situation du Marché de l'emploi dans la Grande Région, Neuvième rapport de l'Observatoire Interrégional du Marché de l'Emploi pour le quatorzième Sommet des Exécutifs de la Grande Région.*

Rumford, C., 2006. Theorizing Borders. *European Journal of Social Theory*, 9(2), pp.155–169.

Schnuer, G., 2014. Circulating in Places and the Spatial Order of Everyday Life. *Human Studies*, 37(4), pp.545–557.

Spierings, B. and van der Velde, M., 2008. Shopping, Borders and Unfamiliarity: Consumer Mobility in Europe. *Tijdschrift voor Economische en Sociale Geografie*, 99(4), pp.497–505.

Spierings, B. and van der Velde, M., 2013. Cross-Border Differences and Unfamiliarity: Shopping Mobility in the Dutch-German Rhine-Waal Euroregion. *European Planning Studies*, 21(1), pp.5–23.

STATEC, 2015a. Travailleurs frontaliers occupés au Luxembourg selon la residence et la nationalité. Available online at: www.statistiques.public.lu/stat/TableViewer/tableView.aspx?ReportId=494&IF_Language=fra&MainTheme=2&FldrName=3&RFPath=92/ [Accessed 7 January 2016].

STATEC, 2015b. Emploi salarié intérieur par lieu de residence et nationalité. Available online at: www.statistiques.public.lu/stat/TableViewer/tableView.aspx?ReportId=7255&IF_Language=fra&MainTheme=2&FldrName=3&RFPath=92 [Accessed 7 January 2016].

Strüver, A., 2005a. *Stories of the 'Boring Border': The Dutch-German Borderscape in People's Minds*. Münster: LitVerlag.

Strüver, A., 2005b. Spheres of transnationalism within the European Union: On open doors, thresholds and drawbridges along the Dutch-German border. *Journal of Ethnic and Migration Studies*, 31(2), pp.323–343.

van Houtum, H., 2000. An Overview of European Geographical Research on Borders and Border Regions. *Journal of Borderland Studies*, 15(1), pp.57–83.

van Houtum, H. and Gielis R., 2006. Elastic Migration: The Case of Dutch Short-Distance Transmigrants in Belgian and German Borderlands. *Tijdschrift voor Economische en Sociale Geografie*, 97(2), pp.195–202.

van Houtum, H. and van der Velde, M., 2004. The power of cross-border labour market immobility. *Tijdschrift voor Economische en Sociale Geografie*, 95(1), pp.100–107.

van Houtum, H. and van Naerssen, T., 2002. Bordering, Ordering and Othering. *Tijdschrift voor Economische en Sociale Geografie*, 93(2), pp.125–136.

5 Crossing territorial borders and social boundaries?

Observations on the German and French workforce in the spa town of Baden-Baden, *c.* 1840–1870

Eva Zimmermann

During a large part of the nineteenth century, Baden-Baden was 'the summer residence of all the nobility and fashion of Europe' (Anon., 1855, p.122). The German spa town, situated about 15 kilometres from the French border in the former Grand Duchy of Baden, had once been the seat of the Margraves of Baden until it was almost completely destroyed by French forces in 1689 during the course of the Nine Years' War and fell into oblivion for the subsequent hundred years. It was only at the time of the Second Congress of Rastatt (1799) that it regained some, if only moderate, international popularity (Steinhauser, 1974, p.100). However, by the mid-nineteenth century, a very different picture comes into view: Baden-Baden had now become the 'Queen of Watering Places' (Anon., 1855, p.122) or, as the French writer Eugène Guinot put it in his then famous work *A Summer at Baden-Baden* – first published in 1847 – the 'summer capital' of Europe (Guinot, 1853, p.8). This career had its beginning in the 1830s, and there followed a long-running boom after 1840 that culminated in the 1860s and came to a sudden end with the Franco-Prussian War in 1870/71.

The nineteenth century, particularly the period from 1840 to 1870, was the Golden Age of European watering places (Gerbod, 1985, p.139). Unlike in the previous century, and notably as a result of the increased mobility that came along with the advent of the railway, recreational travel became less and less a privilege of the nobility. Smaller and greater spa towns mushroomed all over Europe, including in Britain, so that their number exceeded six hundred by the early 1880s (Flechsig, 1883). However, only few reached the rank of what is called a *Weltbad* in German and a *Hydropole* in French: that is, a fashionable spa town of international reputation with an annual attendance of tens of thousands of visitors (Gerbod, 1987, p.312). Among these were several West German towns such as Wiesbaden and Ems; the West-Bohemian spa triangle of Franzensbad, Karlsbad and Marienbad; Vichy and Aix in France; and – already well known in the eighteenth century – Spa in Belgium and Bath in England. In the summer of 1840, Baden-Baden for the first time officially surpassed 20,000 visitors and from then on experienced an unparallelled boom with a regular attendance of over 50,000 from the late 1850s, whereas its competitors levelled off between 20,000 and 30,000 (Frech, 1870; Herrmann, Klemm and Mayer, 2012). The mostly local and regional visitors of earlier times, who had come for medical reasons, were soon marginalized by an international public composed of members of the European aristocracy and the emergent financial, industrial and commercial bourgeoisie, supplemented by celebrities from the world of theatre, music and literature. Between 1840 and 1870 Baden-Baden, which in spite of the growing middle-class attendance at watering places in general remained a rendezvous of the elite, was thus much less of a therapeutic place than a pleasure ground for a privileged leisure class. With up to 50 per cent of all visitors, it also registered the highest rate of foreign guests, who came from all over Europe and, from the 1850s, from other continents as well (Herrmann, Klemm and Mayer, 2012, p.16). However, for more than three decades it was the French who represented a major share of all visitors, surpassed only by guests from the Grand Duchy and other German states, and their influence was thus ubiquitous in Baden-Baden. French was the common tongue of social intercourse that not only 'preside[d] over the cordial understanding of the leading minds of Europe' (Guinot, 1853, pp.183–184), but also served as the main language of commerce and service: 'Every waiter, coachman, artisan and beggar boy smatters French,' whined the German military writer Julius von Wickede (1852, p.190). Moreover, the public space of Baden-Baden was increasingly Frenchified: already by the 1840s most of the hotels had combined German and French names, and the same applied to restaurants and stores (Allgemeine Zeitung, 1860; Wickede, 1852, p.190). Hence, German contemporaries frequently complained that they felt as if they were in a French colony (Andree, 1870, p.92; Wachenhusen, 1863; Wickede, 1852, p.190), whereas Parisian contemporaries tended to take it for 'one of our *faubourgs*' (Pilgrim, 1862, p.272).

This tendency toward Frenchification was fostered by the rapid expansion of the railway network after the foundation of the Grand Duchy of Baden State Railway in 1840 and the French Compagnie des Chemins de Fer de l'Est in 1845. The relevant section of the Rhine Valley line was completed in 1844, followed by the branch line Baden-Baden-Oos in 1845, and the *CF de l'Est* accomplished their main task, the establishment of the Paris-Strasbourg line, in 1852. When France and the Grand Duchy signed an interstate agreement for the construction of the Rhine Bridge at Kehl five years later, the railway became a transnational matter resulting in de-bordering effects. The first permanent railway bridge across the Upper Rhine opened in April 1861 and was particularly celebrated by the Parisian habitués of the spa resort, who could now 'put on an evening dress in the morning in Paris and dance in Baden-Baden by night'[1] (Méry, 1864, p.271). A further reason for the high rate of French visitors, as well as for Baden-Baden's upswing in general, was the gambling halls in the Conversation House. Public gambling was prohibited throughout Europe with the exception of several German spa towns, giving these a crucial competitive edge. Baden-Baden not only had a casino, but also had its leaseholders and brilliant entrepreneurs, the Benazet-Dupressoir family, who acted more as cultural patrons of the town than as mere directors of the green tables. Jacques Benazet had been the principal tenant of several famous Parisian gaming houses for almost twenty years when the French gambling prohibition came into effect in 1838. One year later he came to Baden-Baden, evidently with the intention of turning the small Black Forest spa resort into a little Paris (Rochefort, 1863, p.7). He proceeded very rapidly and effectively in this matter and his successors, son Edouard Benazet and grandson Emile Dupressoir, followed in his footsteps. They hired distinguished French architects and decorators for new buildings and modifications, engaged the companies of the great Parisian theatres, won over the Jockey Club to host the Iffezheim horse races, and did all of this in such a way that it attracted the glamorous society of *Tout-Paris*. 'The brief summer span of life of Baden depended entirely on the Parisians,' declared an English observer in a retrospective written in 1874, three years after the Franco-Prussian War that turned out to be a watershed in the town's history (Anon., 1874, p.302). The case of Baden-Baden thus involves various interacting phenomena that can be grouped under the designation of *de-bordering*, from transnational politico-economic cooperation fostering growing geographical mobility and hence cross-border movement of persons and goods to processes of cultural exchange and knowledge transfer. These tendencies and developments found their focal point not in the whole of a border region, but in a single town located in such a region. In contrast, and as will be discussed below, the implications of the war on Baden-Baden can be seen as symptoms and effects of *re-bordering*.

The great European spa resorts of the nineteenth century have recently been characterized as sites of preservation of a cosmopolitan heritage (Jarrassé, 2012, p.15). Whereas such an assessment primarily refers to the domains of art, letters and architecture, the objective of this study is to broaden the perspective by taking into account two groups of people whose contributions to Baden-Baden's rise, development and maintenance have not yet been considered adequately. The first group to be discussed is the local population and its changing living conditions. A second group of interest is the seasonal workers, notably those coming from abroad. One reason for the extensive neglect of the important role of these groups – considered here as collective actors – may be the available source material, which, notably in the case of the seasonal workers, turns out to be challenging. A basic problem arises from the lack of official statistics: people who came for work were usually not registered in the list of arrivals, so that there are neither reliable figures nor dependable information concerning their country of origin and social status. Furthermore, no ego documents associated with any of the investigated groups have been handed down to us. Finally, the seasonal workers, and to a lesser extent also the local population, have been largely neglected by contemporary commentators. The plentiful reports from Baden-Baden featured in numerous international periodicals – most particularly in French newspapers and journals – barely took notice of them. Some of the rare exceptions are articles by the French-German journalist Wilhelm von Chézy, appearing between 1831 and 1847 in several German newspapers. His reporting, as well as his memoirs published in 1863/1864, differ from the mainstream insofar as they take a look behind the scenes of the fashionable spa society. Further useful data can be obtained from various other sources such as the advertising section of the local gazette, the *Badeblatt*, as well as from travel guides and reports.

Both groups will be examined in terms of their involvement in the dynamics of internationalization that shaped the reputation of Baden-Baden as the summer capital of Europe between the 1830s and 1872. Thereby, it is important to note that both groups not only reacted to the changing conditions of their environment, but also took a proactive role in the process of cosmopolitanization that is to be analysed. Moreover, the two groups will be studied in juxtaposition to illustrate possible dependencies and cooperation. Therefore, the article is structured by business branches, namely the tourism and service industry on the one hand and production and trade on the other.

Tourism and service

By the beginning of the nineteenth century Baden-Baden's traditional industry, which was based on handicraft, had started shifting from the town centre to the outlying districts. The formerly vital tannery industry, for instance, had almost completely disappeared by the beginning of the 1840s (Schreiber, 1840, pp.52–53). This may be an extreme example, and as emerges from the *Baden Directory of Professions and Trades*, published in an English travel guide in 1858 (Coghlan, 1858, pp.124–140), other long-established Black Forest businesses, such as rope making and wood economy, remained relatively important. Nevertheless, the index also reveals a preponderance of the tourism and service industry on the one hand, and the production and trade of consumer and luxury goods, notably clothing, on the other. Hence, the most common male occupation was innkeeper followed by shoemaker and livery stable keeper, while significant parts of the female population worked as laundresses or dressmakers.

Given the repeated assertions about a French dominance in Baden-Baden, it might be surprising to note that the hotel, restaurant and service industry was almost entirely managed and staffed by the local population. Throughout the entire period examined, numerous hotels were established and enlarged, demolished and rebuilt, with a boom in the late 1830s (Herrmann, Klemm and Mayer, 2012, p.69). Contemporary travel guides list up to fifty guesthouses, about twenty of which were large hotels and between ten and twelve rated as first-class (Coghlan, 1858; Joanne, 1863, p.121). The family names of the hoteliers as well as the little additional information available suggest that they were mostly natives. Sometimes their families had been working in the gastronomy sector for several generations. Take for instance the Kah family, whose name appears time and again among the hotel and restaurant keepers listed in the *Baden Directory*. It appears that many of the fathers of the workers in this sector, and sometimes the workers themselves, had been artisans and craftsmen before changing to tourism (Wettstein, [no date]). Such was the case with Ignaz Stadelhofer, whose career might serve as an example here: discerning the signs of the times, he and his wife founded the luxurious Hôtel d'Angleterre in the immediate vicinity of the Conversation House in 1836.

Stadelhofer thus moved on from life as a cooper to an international reputation as a 'worthy host [...] famed far and wide for his urbanity' (Anon., 1855, p.125). In fact, innkeeping in a place like Baden-Baden involved a considerable degree of individual cosmopolitanism, concisely described by the Scottish travel writer John MacGregor in 1837 using the example of the Hôtel de Fortune in Offenburg, a mere 50 kilometres away from Baden-Baden:

> By the way I must note down that the Innkeepers in Germany [...] are usually persons of very extensive information. Who possesses and keeps the Fortune, not only speaks English, French, and his native language well, but on most points of useful knowledge I found him much more intelligent than many members of both houses of the British Parliament with whom I have casually been in contact. On rents, revenues, soils, minerals, and the laws of all the German states, and of general subjects relative to France, Switzerland, and other countries, the very obliging landlord I speak of was remarkably, yet modestly, conversant.
>
> (MacGregor, 1837, pp.1–2)

While smaller inns may have been conducted as family enterprises, the large first-rate hotels were certainly in need of additional staff. Several sources tell of the 're-migration' of a crowd of seasonal workers, namely 'waiters, cooks, maids and servants, all dismissed for the winter and coming mostly from Alsace'[2] (Anon., 1847, p.470), whose annual arrival in late April announced the beginning of the season. Unfortunately, there are no appropriate sources providing further information about hotel staff such as porters, bell-boys or chamber maids. The often reported fact that servants used to speak French by default and that their appearance seemed to identify them as French as well (Morgenblatt, 1846b, p.610) is no proof of their actual national origin. Like their employers, they usually spoke several tongues and notably the mastery of the French language was indispensable in Baden-Baden as in other German watering places. Thus a German-French dictionary published in 1857 under the telling title *Der fertig französisch sprechende Kellner* (*The Perfect French-speaking Waiter*) was especially addressed to 'inn-keepers, waiters and inhabitants of spa towns' (Anon., 1857). As for the waiters and cooks, several sources confirm that many of them were effectively French (Anon., 1840, p.206; Lallemand, 1858, p.10; Steub, 1873, p.255).

However, the capacity of the hotels did not suffice to accommodate the ever-increasing influx of summer guests. As a result, the chief business of the local population – not mentioned in the *Baden Directory* – consisted in letting furnished lodgings. By the end of the 1850s, 390 of the 619 houses of the town served as lodging houses, starting from small apartments for the less wealthy right up to entire houses with numerous bedrooms, several saloons, kitchens, stabling and a coach-house for those noble guests who wished to have homelike privacy (Coghlan, 1858, pp.27–28). 'As soon as the month of May arrives, the population of Baden-Baden disappears to make room for strangers. It is impossible to say where the six thousand inhabitants of the town are lodged,' remarked Eugène Guinot (1853, p.12). The weeks before their 'disappearance' were devoted to the maintenance, enlargement and embellishment of the houses (Lallemand, 1858, p.10). Chézy, who besides his occupation as a journalist operated as a lessor during summer, mentioned in his memoirs that many natives of Baden-Baden borrowed heavily from Swiss banks in order to answer the demand and to satisfy their guests' special needs (Chézy, 1864, p.101). When the guests finally arrived the landlords and their families did not actually disappear, but provided them with goods of daily need and service and, at times, let them take part in their daily life. This could lead to close contact between the hosts and their guests, as in the case of Chézy himself, who reported that, as a landlord, he 'came in closer touch with people and spheres, of whose goings-on he had otherwise rarely or never heard'.[3] Apparently the locals of Baden-Baden did not care much about the criticism aimed at them by parts of the German press for servility, lack of patriotism and the loss of regional identity (Wickede, 1852, p.190; Riehl, 1854, pp.261–262), which they and others considered as hospitality and cosmopolitanism (Crocius, 1859, p.81). In fact, the locals continuously fought against the ever-impending closure of the gaming house, which they deemed rightly to be one of the main causes of the town's success (Gemeinderat der Stadt Baden-Baden, 1862).

The labour market for domestic service in Baden-Baden was not limited to the hotel and private letting industry. Upper-class guests were usually accompanied by at least some of their servants, but others were hired on site. Here the 'cohorts of domestics' (Lallemand, 1858, p.10) were to be found. Applications for posts as private cooks, chambermaids, messengers, nursemaids, tutors or lady's companions were published in every issue of the local gazette, the *Badeblatt für die großherzogliche Stadt Baden* (1829–1882), whose editorial office also served as a placement agency. Coghlan (1858, p.29) specifies the wages of the servants, which ranged from 20 florins a year for a 'housemaid or servant of all work' up to 300 florins for a 'man servant'. A sample review of the mentioned applications confirms existing research on nineteenth-century domestic service work (Fauve-Chamoux and Fialovà, 1997): they originated in general from young people, a majority of whom were female, probably unmarried, and in the present case often from France and most notably from Alsace. Although no reliable figures exist, it can be assumed that domestic servants, hired on site or brought by the guests, represented a significant portion of the foreigners in Baden-Baden. The French historian Paul Gerbod assumes that this applied to the great European spa resorts in general (Gerbod, 1987, p.315). The high attendance of service workers in Baden-Baden is furthermore confirmed by the fact that by 1837 a savings bank had been founded that particularly addressed its services to domestic service workers, and explicitly to the foreign ones as well (Schreiber, 1840, p.50).

Much in contrast to the sometimes excessive reporting about the high society's idle life in Baden-Baden, very little is known about the domestics' living conditions and environment. At least some insight into their daily life is delivered by Wilhelm von Chézy, and also by his colleague Dudley Costello. This Anglo-Irish writer asserted that 'it is one of the great charms of Baden-Baden that servants can enjoy themselves as much as their masters' (Costello, 1847, p.322). In reality, nevertheless, they stayed among their own class and were excluded from many activities and places, for instance from gambling, and their presence at the fashionable promenade in front of the Conversation House was rather unwanted (Morgenblatt, 1844, p.816). However, they had their own mealtimes, organized balls and reunions, and it appears as if the atmosphere at these gatherings of people of all nations, ages and of either gender was often much more at ease than at the social occasions frequented by the higher classes (Morgenblatt, 1825, p.784).

Production and trade

As stated above, the production as well as trade of consumer and luxury goods, notably clothing, shoes and accessories, was the second most important industry in Baden-Baden during the period under investigation. While the market for consumer products, such as pastries, meat, beer, wine and tobacco, was saturated with local and regional products, a different situation seems to have existed in the luxury goods industry. The aforementioned fashionable promenade was not only the central outdoor venue for social interaction, but also an important and thus contested place for trade. As the German-speaking territories had no metropolis similar to Paris or London, the major and minor fairs and the great spa resorts were the only centres of international trade (Mohl, 1845, p.113). Indeed, some contemporary descriptions of the Baden-Baden promenade evoke the atmosphere of an international bazaar:

> Quadruple lines of chestnut trees encompass the grass-plat [sic] on three sides and form as many shaded and well-frequented walks, along the exterior of which ranges of little Bazaar shops, or light *boutiques,* have been established, filled with gewgaws of all sorts and from all parts of the world, served out by smartly dressed, good-looking young women, clad in the costume of their respective countries.
>
> (Granville, 1837, p.43)

There were twenty-three wooden booths which were rebuilt in stone in 1868 on the initiative of Edoaurd Benazet, who was contractually obliged to such conversions and embellishments. It is true that the promenade was an emporium for the trade of international goods, from Bohemian glassware to Italian jewellery and Tibetan cashmere (Lindsay, 1838, p.9). But it is also true that, at least in the later part of the period examined, the majority of the merchants were natives from the Grand Duchy and mostly from Baden-Baden itself, supplemented by one or two fashion retailers from Paris and a few importers from the Austro-Hungarian territories, such as milliners and glovers from Tyrol, glaziers from Bohemia and dressmakers from Hungary. This is confirmed not only by the list of merchants published in the *Baden Directory*, but also by the available documents of the municipality concerning booth leases between 1862 and 1866 (Badeanstalten Commission, 1854–1869). Yet already in 1838, *Fraser's Magazine* had published some letters from Baden-Baden in one of which the author attempted to open the readers' eyes in regard to the young women 'from all parts of the world', who he identified as 'young woman of good old Baden-Baden, blood and bones, though tricked out in different sorts of frippery corresponding to the costumes of the countries from whence their stocks in trade are supposed to come' (O'Donoghue, 1838, p.148). However, it appears as if there was an interim development to the disadvantage of the local mercantile community in the 1840s. This is at least suggested by the fact that in 1846 several native merchants joined forces and appealed to the responsible municipal authority with a request to supply at least a part of the market at the promenade with their own products. Their concern was heeded in the subsequent year when six of the boutiques were provided to local dealers (Morgenblatt, 1847, pp.613–614; 619–620). Moreover, the ministry in Karlsruhe, taking note of the economic potential of the promenade, rented one of the booths in a prime location at public expense in 1851, with the objective to promote local craftwork, namely the Black Forest wood carving industry (Austria, 1851, p.545). This example shows how the locals' open-mindedness towards foreign and notably French influence and impact was nevertheless supplemented by the will not only to maintain their regional identity but also to make it an active part in the process of internationalization.

The locals' success came at the expense of foreign traders, who had less and less chance to obtain the lease for one of the booths and were in a small minority from at least the end of the 1850s onwards. The situation in the adjoining streets did not differ much from that in the promenade. There was simply not enough space for foreign traders to establish new businesses and shops. A compromise in which both parties won was found by at least some merchants: sale on commission. The numerous advertisements in the *Badeblatt*, originating mainly from female fashion retailers from Paris and Strasbourg who sold their items on consignment in local craftsmen's shops, prove that this form of cross-border cooperation was often employed. At times this could result in odd configurations, as in the case of a Parisian dressmaker who placed her stock at the local tinsmith's store. Yet, there were also examples of more reasonable collaboration that could lead to further cooperation. For instance, a native tailor concluded a shareholders' agreement with his Parisian colleague Chevard in 1846. As Chezy reported, the model turned out to be very successful for both sides and soon found many imitators (Morgenblatt, 1847, p.619). Several of the merchants who consigned their goods to locals were themselves present in Baden-Baden, at least for parts of the season. They were joined by the employees of the Conversation House, notably croupiers, and other craftsmen and free professionals whose advertisements can also be found in the *Badeblatt*. Like the dressmakers they came mostly from France, but in this case they were mainly male. They included language teachers, dancing and fencing masters, hairdressers and a crowd of portrait painters, who in later years received competition from photographers. Just like the foreign domestic service workers who 're-migrated' to Baden-Baden as soon as the season began, the merchants and free professionals used to return every year, so that by and by they became somewhat local themselves.

The last days of October annually announced the end of the 'eternal feast', as the summer season in Baden-Baden was often described (Mornand and Vilbort, 1863, p.118). Time and again the local population made efforts to animate the town in winter, but these remained without substantial results (Morgenblatt, 1846a, pp.69–71). 'In wintertime, Baden-Baden doesn't exist anymore,'[4] declared Louis de Cormenin as late as 1868 (p.184). Having been the first migrant birds in late April, the seasonal workers were also the last to leave. During autumn and winter, many inhabitants of Baden-Baden saw themselves as deprived of their primary income source. Some of them lived on their earnings from the past season while others found occupation in the construction business, which – owing particularly to the intense building activities funded by the Benazets – was an important winter industry (Loiseau, 1861, p.25). If Baden-Baden died in the cold season, it resurrected itself even more glamorously in the following year. It was the Franco-Prussian War of 1870/1871 that finally interrupted this seemingly never-ending cycle and forcefully suspended the dynamics of internationalization. 'Baden-Baden is dead,' announced a French journalist in 1873 (La Presse, 1873), and this time it was for good. Before taking a final view on the short- and middle-term consequences of the war, the results of the present study will be briefly summarized.

Synopsis

Regarding the local population of Baden-Baden during its Golden Age, it has been shown that they were willing and prepared to change their lives significantly in order to adapt themselves in various aspects to shifting economic and social conditions. However, 'adaptation' is a somewhat misleading term since reaction and initiative went hand in hand here. Thus, many natives managed to profit significantly from the opportunity of upward social mobility that came along with their home town's international success. The group most likely to benefit were the innkeepers, who for some appeared to be 'the lords of the place' (Granville, 1841, p.72). Some of them, such as Ignaz Stadelhofer, came into close contact with the European nobility of birth, wealth and merit, and, as the example of Chézy has shown, the same applied for private landlords. Moreover, referring to adaptation involves the risk of overlooking the proactive role taken by the local population in Baden-Baden's transformation from a barely known watering place into the summer capital of Europe. Apart from providing the necessary infrastructure, goods and services, they also contributed to the development of sociability. A striking example is Heinrich Haug who gained a 'European name' not only as the innkeeper of the Restauration in the Conversation House, but also for hosting the last international event of each season, the autumnal hunts (Morgenblatt, 1848, p.225). There is no doubt that the local population proved to be very open-minded regarding foreign culture and conventions – notably those of their main clientele, the French.

Yet it is myopic only to consider the dynamics of Baden-Baden as a one-sided French-German transfer, since one should not omit the locals' function as agents of cultural exchange. Such exchange happened, for instance, by promoting regional craftwork and consumer goods, like wine and tobacco, or simply by conversing and sometimes sharing part of their daily lives with the guests. The locals also adhered to their traditions and values, like celebrating their feasts, such as the anniversary of the constitution, but also, on 24 August – and thus in presence of the numerous French summer guests – the commemoration of the Great Fire caused by French soldiers in 1689. Other occasions were commonly celebrated, notably the Grand Duke's birthday, where the festivities were supported by the prudent Benazets who were well aware of the importance of the local population for their enterprise. In summary, what patriotic German commentators censured as too much willingness to assimilate, can also be seen as a tendency towards collective and individual cosmopolitanism, which was thus neither a peculiarity nor an intellectual privilege of the elites. Furthermore, such behaviour did not necessarily imply a loss of regional identity, but rather introduced elements of the latter into the process of cosmopolitanization.

Regarding the different groups of seasonal workers from abroad, it has been shown that, just as with the guests, they primarily came from France. One of the main reasons for this tendency – in addition to the close border and the ever-improving travel opportunities between the two countries – was the Frenchification of public space, including the language, that was supported by large parts of the local population. In contrast to the permanent inhabitants of Baden-Baden, the seasonal workers' decision to be a part of the development and maintenance of Europe's summer capital was more or less deliberate. At the same time, however, they had less immediate influence on changes and developments than the former, and thus experienced more pressure to adapt to the conditions they found. Nevertheless, they profited from the open-mindedness of the local population no less than the international guests and were thus enabled to overcome various barriers such as language or trade restrictions, in such way achieving the degree of geographic mobility important for them. In contrast, upward social mobility may have been comparatively rare in their spheres – at least no such case is reported.

Concerning the foreign domestics', traders' and free professionals' daily routines and social life in Baden-Baden, the contemporary commentators' quasi-ignorance toward all of them impedes deeper evaluation. Yet, the few existing references to their free time activities suggest that a form of transnational sociability prevailed that was more liberal in practice and in attitude than the sometimes pretentious cosmopolitanism of the upper classes. Moreover, like the celebrated international musicians and actors who graced the town with their presence, the foreign seasonal workers acted as agents of cultural transfer, and one cannot underestimate their impact on the dynamics of cosmopolitanization, which not only involved high culture but also everyday life, such as eating and drinking, leisure activities like dancing and fencing and not least fashion.

The various bridges built between the local population of Baden-Baden, foreign guests and seasonal workers were, however, no more lasting than the Rhine Bridge at Kehl, which was demolished almost immediately after the French declaration of war on 22 July 1870. The town and its inhabitants experienced a sort of shock, as reported by the English periodical *Harper's Bazaar* on 17 September:

> An hotel-keeper in Baden-Baden writes that war is ruinous to all there. It broke out just as Baden began to be full. Now, he says, 'we have not a soul staying with us in the hotel. My servants are all gone, either to the army or home. There are no flowers now on our staircases and balconies, and everything is shut up, just as if one were in the middle of winter.'
>
> (Harper's Bazaar, 1870, p.599)

The outbreak of war turned out to be a permanent cessation in the process of Baden-Baden's cosmopolitanization, which now had to face the consequences of the political, cultural and mental re-bordering between France and Germany. The situation was further worsened by the German gambling prohibition finally achieved by the Imperial government in 1872. The town's last great impresario, Emile Dupressoir, was certainly not the only Frenchman to turn his back on the place. 'Once, it was declared, without appeal, that to go to Germany was ostracism for life for every Frenchman, this gay little "Opéra Comique" village was destined to fall as it has fallen,' commented the author of the Baden-Baden retrospective quoted in the beginning (Anon., 1874, p.203). In fact, the absence of the French guests dramatically affected the visitor statistics as well as the mode of life and culture. Although the town managed to recover numerically – registering a growth of the total number of guests within the next three decades (Herrmann, Klemm and Mayer, 2012, p.42) – it has never been able to regain its reputation as the summer capital of Europe. The number of foreign visitors, particularly from France and England, once the second most important nation, remained at a very low level, and by the end of the century foreign cities were not to be found among the most frequent points of origin anymore (Herrmann, Klemm and Mayer, 2012, p.42). By focusing on its medical function as a therapeutic bath (Eidloth, 2012, p.71), Baden-Baden now primarily attracted a regional middle-class attendance – and thus somewhat returned to what it had been a century before.

The long-term effects on the examined groups is a subject for further investigation, but some hypotheses can be ventured. Regarding the foreign seasonal workers, it seems likely that their number decreased in parallel to the ebbing of international upper-class attendance. Notably the French traders, free professionals and domestic service workers found themselves left without their main clientele, the society of *Tout-Paris.* As for the local population, the transformation of Baden-Baden did not necessarily affect their economic livelihood since, as stated above, the total number of visitors was showing an upward trend. However, one may nevertheless conjecture that the loss of internationality and cosmopolitanism became quite noticeable in their cultural, mental and social lives after the summer capital of Europe had come to an end.

Notes

1 "On revêt le matin un costume de bal à Paris, et on danse le soir à Bade."

2 "In ganzen Schaaren kommen Kellner, Köche, Mädchen und Knechte, für den Winter alle entlassen, größtentheils aus dem nahen Elsaß, wo die Meisten derselben ihre Heimath haben, wieder eingewandert."

3 "Die Zeit, welche Chézy auf dem ehemaligen „Kornfrieder" als Hausherr verlebte, hat ihn mancherlei Menschen und Kreisen in nähere Verbindung gebracht, von denen [sic!] Thun und Treiben er selbst wenig oder nichts erfahren hätte."

4 "En hiver, Baden[-Baden] n'existe plus."

Bibliography

Andree, K., 1870. Unsere deutschen Grenzen und unsere Nachbarn. *Globus: Illustrirte Zeitschrift für Länder- und Völkerkunde*, 18, pp.90–93.

Allgemeine Zeitung, 1860. Kissingen und Baden-Baden (Schluss). *Allgemeine Zeitung*, 8 Aug. pp.3678–3679.

Anon., 1840. Baden-Baden und die Spielbank. *Deutsche Vierteljahrs Schrift*, 1, pp.204–224.

Anon., 1847. Heidelberg und Baden-Baden. *Die Grenzboten*, 6(3), pp.468–479.

Anon., 1855. Home Life in Germany. *The Illustrated London Magazine*, 5, pp.117–126.

Anon., 1857. *Der fertig französisch sprechende Kellner.* Nürnberg.

Anon., 1874. Baden: A Memory. *London Society*, April 1874, pp.302–312.

Austria, 1851. Aus Südwestdeutschland. *Austria: Tagblatt für Handel und Gewerbe, öffentliche Bauten und Verkehrsmittel*, 10 April, p.545.

Badeanstalten Commission [Großherzoglich Badisches Bezirksamt Baden], 1854–1869. *Die Handelsbuden auf der Promenade in Baden, deren Verpachtung, Unterhaltung etc.* Generallandesarchiv Karlsruhe. 236/3360. Karlsruhe.

Badeblatt für die großherzogliche Stadt Baden, 1829–1882. Baden-Baden.

Chézy, W., 1864. *Erinnerungen aus meinem Leben, vol. 2(3): Helle und dunkle Zeitgenossen.* Schaffhausen.

Coghlan, C. F., 1858. *The Beauties of Baden-Baden and its Environs.* London.

Costello, D., 1847. Adrien Roux; or, the Adventures of a Courier. *The New Monthly Magazine and Humorist*, 2, pp.321–337.

Crocius, P., 1859. Chronique. *Illustration de Bade*, 29 Jul., pp.80–81.

de Cormenin, L., 1868. *Reliquiae*, vol. 2., Paris.

Eidloth, V., ed., 2012. *Europäische Kurstädte und Modebäder des 19. Jahrhunderts.* Stuttgart: Konrad Theiss.

Fauve-Chamoux, A. and Fialova, A., eds., 1997. *Le phénomène de la domesticité en Europe, XVIe – XXe siècles.* Praha: Ceská Demografická Sociologicka, Acte Demographica, 13.

Flechsig, R., 1883. *Bäder-Lexicon.* Leipzig.

Frech, C., 1870. *Der Kurort Baden-Baden.* Karlsruhe.

Gemeinderat der Stadt Baden-Baden, 1862. *Petition der Stadt Baden, die Aufhebung des öffentlichen Spieles daselbst betreffend.* Stadtarchiv Baden-Baden. 02-162/008. Baden-Baden.

Gerbod, P., 1985. Le loisir aristocratique dans les villes d'eaux françaises et allemandes au XIXe siècle (1840–1870). *Pariser Historische Studien*, 21, pp.139–154.

Gerbod, P., 1987. Les 'fièvres thermales' en France au XIXe siècle. *Revue Historique*, 277(2), pp.309–334.

Granville, A. B., 1837. *The Spas of Germany*, vol. 1. London.

Granville, A. B., 1841. *The Spas of England, and Principal Sea-Bathing Places.* London.

Guinot, E., 1853. *A Summer at Baden-Baden.* London.

Harper's Bazaar, 1870. Sayings and doings. *Harper's Bazaar*, 17 Sept., p.599.

Herrmann, K., Klemm, T. and Mayer, M., 2012. *Internationalität in augewählten Kurstädten des 19. Jahrhunderts.* Baden-Baden.

Jarrassé, D., 2012. *Gutachten über die Vorzüge der Stadt Homburg vor der Höhe (Hessen) zur Aufnahme in die Welterbeliste im Rahmen einer Gruppenbewerbung unter dem Namen Great Spas of Europe.*

Joanne, A., 1863. *Bade et la Forêt-Noire.* Paris.

La Presse, 1873. Chronique. *La Presse*, 6 Jul.

Lallemand, C., 1858. Petite Chronique de Bade. *Illustration de Bade*, 22 May, p.10.

Lindsay, E., 1838. *Le Grand-Duché de Bade.* Trans. Demoulins, E.A. Paris.

Loiseau, L., 1861. Murmures de l'Oos. *Illustration de Bade*, 17 Jun, pp.25–27.

MacGregor, J., 1837. *My Notebook: Switzerland.* Frankfurt o. M.

Méry, J., 1864. *Les Amours des bords du Rhin.* Paris.

Mohl, M., 1845. *Aus den gewerbswissenschaftlichen Ergebnissen einer Reise in Frankreich.* Stuttgart.

Morgenblatt für gebildete Stände, 1825. [Korrespondenz-Nachrichten: Baden bei Rastatt]. *Morgenblatt für gebildete Stände*, 16 Aug., pp.779–780.

Morgenblatt für gebildete Leser, 1844. Die Saison. [Korrespondenz-Nachrichten: Baden-Baden, August]. *Morgenblatt für gebildete Leser*, 24 Aug, pp.815–816.

Morgenblatt für gebildete Leser, 1846a. Ein Badeort im Winter. *Morgenblatt für gebildete Leser*, 21 Jan, pp.69–71.

Morgenblatt für gebildete Leser, 1846b. Badener Zustände. *Morgenblatt für gebildete Leser*, 27 Jul, pp.609–610.

Morgenblatt für gebildete Leser, 1847. Badener Zustände. *Morgenblatt für gebildete Leser*, 29 June pp.613–614; 30 June, pp.619–620.

Morgenblatt für gebildete Leser, 1848. Heinrich Haug zum Salmen. *Morgenblatt für gebildete Leser*, 7 March, pp.225–227.

Mornand, F. and Vilbort, J., 1863. *Voyage illustré dans les deux mondes.* Paris.

O'Donoghue, C., 1838. Ensign O'Donoghue's Packet from Belgium. *Fraser's Magazine for Town and Country*, 18, pp.140–155.

Pilgrim, L. [pseud.], 1862. Les stations de la villégiature. *L'Artiste : Beaux-Arts et Belles-Lettres*, 32(1), pp.270–273.

Riehl, W. H., 1854. *Die Naturgeschichte des Volkes als Grundlage einer deutschen Social-Politik, vol. 1: Land und Leute.* Stuttgart.

Rochefort, H., 1863. Bade en 1863. *Le Journal amusant*, 19 Sept. pp.6–7.

Schreiber, H., 1840. *Baden-Baden: die Stadt, ihre Heilquellen und Umgebung.* Stuttgart.

Steinhauser, M., 1974. Das europäische Modebad des 19. Jahrhunderts. In L. Grote, ed., 1974. *Die deutsche Stadt im 19. Jahrhundert.* München: Prestel, pp.95–128.

Steub, L., 1873. *Kleinere Schriften*, vol 1. Stuttgart.
Wachenhusen, H., 1863. Wachenhusen über das deutsche Badeleben. *Diadaskalia*, 19 July, [no pagination].
Wettstein, R., [No date]. *Traditionsreiche Hotes in Baden-Baden.* Available online at: www.bad-bad.de/gesch/trad_hotels.htm [Accessed 11 October 2015].
Wickede, J., 1852. *Aus dem Leben eines Touristen.* Altona.

6 Crossing borders

Politico-geographical and mental borders in contemporary German-language literature in Belgium

Lesley Penné

'Land without borders'. This is the title of the poem with which Leo Wintgens begins the first part of his novel *Wege aus Sümpfen: Roman einer Grenzlandschaft* (Paths Out of the Moor: Novel of a Border Landscape).[1]

land ohne
grenzen
pays sans
frontières
land zonder
grenzen

«Land
zonder grenzen»
een slag
woord voor toeristen
die even
de streek
bewonderen
van buiten

«Pays
sans frontières»
de barrières
il n'y en a plus –
ne restent
que les anciens octrois
que l'indigène
qui ressent toujours
mieux les frontières
doit passer
comme partout et toujours

land
ohne
grenzen ein SCHLAG
nur auf
die alte hohle pauke
der manche zänkische
geschichtlich genährte
vorurteile
im kontrapunkt untermalt

Figure 6.1 Poem: 'Land without borders.'
Source: Wintgens (2001).[2]

The fact that the author publishes this meaningful poem, rich with metaphors, on the first page of his novel demonstrates clearly the importance that is attributed to the notion of 'borders'. Wintgens is not alone in doing this. Other German-language authors in Belgium have also given a central role to borders in their novels. In this chapter I will discuss the different kinds of borders and their meanings in three German-language Belgian novels: *Wege aus Sümpfen. Roman einer Grenzlandschaft. Teil I: Eine Jugend auf dem Königshof* (2001) by Leo Wintgens, *Bosch in Belgien* (2006) by Freddy Derwahl, and *Unterwegs zu Melusine* (2006) by Hannes Anderer.[3] What kinds of borders appear in these novels? How do the characters experience these borders, and what is their literary function?

The first reason for choosing these three novels rests on the fact that they were all published around the turn of the millennium by authors who belong to roughly the same generation. Second, their histories are all situated in the same spatial and historical context and narrated by a male protagonist who was born during or shortly after the Second World War. Third, they are representative of a shift in the form of remembering that gained importance in the last decades. Instead of trying to forget the past by staying quiet about it, this generation wants to remember the past in order to learn to deal with it, and to try to move on rather than stay stuck with untreated traumas. Literature can be a useful medium in this process. Inspired by the literary theory of Aleida Assmann on cultural memory, I will try to provide an insight into the relationship between the concepts of borders, *Heimat*, the past and identity in contemporary German-speaking Belgian literature.

Historical and geographical contextualization

To understand the place of borders in Belgium's German-language literature, an understanding of the geographical situation and the history of this region is necessary.[4] In addition to the Dutch- and French-speaking parts of Belgium, there is also a small German-speaking part in the east of the country, with about 76,000 inhabitants. The German-speaking Community is located in the province of Liège in Wallonia and is administratively divided into the municipality of Eupen in the north, situated at the border with Germany and touching the Netherlands at one point, and the Belgian Eifel in the south, on the border with Luxemburg and Germany. The German-speaking Community of Belgium has its own parliament and government in Eupen. It was in the eighteenth century that the region first became part of the same jurisdiction, namely of the French *département* of Ourthe. After the defeat of Napoleon it was assigned at the Congress of Vienna to the Kingdom of Prussia, and German became the official language. The region became part of Belgium in 1920 after the First World War, as a condition of the Treaty of Versailles.

In 1940 it was annexed to the German Reich, and in 1945 it returned to Belgium. In the post-war period the Belgian government undertook a large-scale purge that was conducted much more severely in the German part than in the Flemish or Walloon parts of Belgium. It not only led to the suppression of the German language and the local Germanic dialect in favour of the French language, especially in administration and education, but the German-speaking population was also subjected to a special law that made it possible to deprive them relatively easily of their Belgian nationality. There was a strong atmosphere of suspicion among the French- and Dutch-speaking Belgian population towards the German-speaking population, and there was widespread disinclination to consider the specificity of the situation in the border region that resulted from the German annexation (Sepp, 2010, p.14).

As for the members of the German-speaking population, they were left with a sense of homelessness. Their attitude towards Belgium was a reserved one, to say the least, and thinking of Germany as their *Heimat* was no longer an option. It was only in 1963 that the Belgian government recognized the German language again as an official language in education and administration. With the first state reform in Belgium in 1970, three cultural communities were established: the Dutch Cultural Community, the French Cultural Community and the German Cultural Community. This was the first time since the Second World War that the German Cultural Community had cultural autonomy. With the second state reform in 1980, the names of the communities changed to the Flemish Community, the French Community and the German-speaking Community.

The German-speaking Community of Belgium had a turbulent history in the past century, with a triple nationality switch over a period of just twenty-five years. This transition between two countries and the accusations – partly true and partly unjust – of collaboration with the German Reich in particular, led to the problematic image that the German-speaking population of Belgium had (and in part still has) about its own past. According to Aleida Assmann there are two different ways to remember the past: one is by the 'preservation' of the past (*Vergangenheitsbewahrung*); the other is by 'coming to terms' with the past (*Vergangenheitsbewältigung*). Remembering by preserving the past implies permanently bearing the responsibility of the past. On the other hand, remembering by coming to terms with the past implies an engagement with the past and remembering those things that one would prefer to forget in order to move on and enable social integration (Assmann, 2013, pp.115–116).

Although it is quite difficult for the German-speaking Belgian population to find an unproblematic relationship with the past, the will to engage with their regional history has grown in the last decades (Fickers, 2004). Literature is one element in this difficult process of remembering. It is the second form of remembering – coming to terms with the past – which the authors of the novels discussed here try to achieve. Through family histories they engage in and rewrite the history of their home region. Focusing on the topic of the border they explore the regional past, and in this way the border transcends its narrow spatial meaning and develops into a ubiquitous phenomenon that goes beyond clear geographical and politico-historical demarcations. In this form, the concept of the border becomes an integral part of East Belgium's cultural memory (Beck, 2010, p.135).

Opening borders

To begin the discussion on borders in the German-language literature of Belgium, we can examine the poem printed above, as it brings together several important insights concerning the topic. This poem includes all the key dimensions of borders in Belgium's German-language literature: politico-geographical borders, mental borders, the difficult history, and language issues.

What immediately attracts the attention of the reader is the fact that the poem is written in three different languages. It consists of three parts with different contents in each, written in Dutch, French and German respectively. The Dutch part is that of the tourist who admires the region from the outside. It refers to the enjoyment a number of Flemish people take in the beautiful nature of the region, and to their penchant for winter sports there, i.e. to the mostly recreational and somewhat superficial relation they maintain with this part of Belgium. The French strophe is about ancient impositions, i.e. administration. The language choice is not surprising as the German-speaking Community of Belgium depends, from an administrative point of view, on the Walloon region of the country. The German part features historically-fed prejudices, which are indeed based on the fact that the region shares a common language with Germany, the country of the enemy during the two world wars.

When we take a closer look at the text we notice the play that is set up between different kinds of borders – politico-geographical borders on the one hand and a multiplicity of mental borders on the other. Mental borders are defined here as constructions that exist in people's minds and that, in many cases, are not easily overcome by the gradual disappearance of political borders.[5]

Each of the three languages adds a different dimension to the possible significance of the expression 'land without borders'. The Dutch beginning of the poem, with the tourists admiring the region from outside, focuses on the geographical and regional borders. The meaning of the word 'outside' (*buiten*) is accentuated by the word's graphic position in the verse, as it is placed far to the right and thus is literally outside the poem. The question that arises, though, is where this 'outside' is situated, as the title of the poem suggests that there are no (more) borders. It seems that the second, French part of the poem gives an answer to this question by adding a new, mental dimension to the significance of 'border'. The point of view is now no longer that of the tourist, but has been shifted to that of the inhabitant of the region himself, who 'feels' (*ressent*) the borders better and who has to cross the ancient impositions of anywhere anytime. So the perspective not only changes from an external to an internal point of view, but this part also opens the borders in the sense that it generalizes the specific regional situation with the words 'like anywhere and anytime' (*comme partout et toujours*).

Presenting the German-language region in Belgium as an example for other regions and countries that should transcend their borders and think more broadly is commonly done in the German-language literature in Belgium. Interesting in this respect, though, is the word *doit* in the poem, which means 'must'. The native (*l'indigène*) is obliged to cross the borders, probably because he does not have any other option if he wants to move forward. While the visual borders have disappeared, he continues to be aware of the borders. These mental borders keep the native captured and prevent him from thinking wider and further. Consequently, transgressing the borders is necessary to open up a future perspective. It therefore is not surprising that in the last, German part the historical dimension of the 'land without borders' – already partially present in the second part with the barriers that have disappeared – has been placed at the centre of our attention. The author achieves this by adding acoustic associations that accentuate this time-related, historical dimension. He mentions the historically-fed prejudices that arise by a strike of the kettledrum, which is old and hollow. By the use of the word 'strike' ('SCHLAG'), which is emphasized by capitals, and the word 'kettledrum' (*pauke*), the author creates a military atmosphere. This is reinforced by the use of the German language, which was the language of the enemy during the two world wars.

This '*strike*' (*SCHLAG*) also reminds us of the word *slag* (stroke) in the first, Dutch part of the poem. This is a very nice wordplay by the author, as the Dutch word '*slag*' is followed by '*woord*' (word). '*Slagwoord*' can mean either a 'slogan' or a 'word that busts into something'. In the Dutch and French parts, 'land without borders' is indeed still a slogan and is to be understood figuratively. The Dutch and French populations of Belgium do see the region as a land that makes the connection between two countries, but for them the concept of border and the slogan 'land without borders' do not have the same profound historical meaning that they do for the German-speaking population. The meaning of a slogan disappears in the German part, where the word '*SCHLAG*', reinforced by the use of capital letters, appears alone, without reference to the word '*slagwoord*'. For the German-speaking population 'land without borders' is not a slogan, but has to be understood in a factual sense: the borders have disappeared. The words 'land without borders' bust into the kettledrums, but have also bust into the mind and the memory of the people. It is a real strike, loaded with emotions. The kettledrum is old and hollow, and as such it acts as an accompanying counterpoint to ancient quarrelsome prejudices. The voice of the region thus accompanies the 'sound' of historically nourished prejudices and offers another perspective on the past.

It is also important to note that the quotation marks of 'land without borders' have disappeared in the German part. This reinforces the hypothesis that the notion of slogan has disappeared here and can moreover be interpreted as another, graphic, border that has vanished. The last part of the poem seems to indicate that the region – the land without borders – managed to give an answer to ancient prejudices and has liberated itself from the borders. By thus speaking of the *Vergangenheitsbewältigung*, the coming to terms with the past that the German-speaking population longs for, the poem is, so to speak, emblematic of the work of the three authors to be discussed.

Liberation from borders

With the aforementioned key elements in mind – politico-geographical borders, mental borders, the difficult history and language issues – we can now take a closer look at the way in which the concept of the border has been used and developed in the German-language Belgian novels mentioned above: *Wege aus Sümpfen. Teil I* by Leo Wintgens, *Bosch in Belgien* by Freddy Derwahl and *Unterwegs zu Melusine* by Hannes Anderer. The three authors belong more or less to the same generation: those who were in their early childhood during the Second World War or who were born immediately after the war. The three novels deal with similar themes and share the same geographical and historical setting. All three novels feature a male protagonist who is born shortly after the Second World War, and so the three plots are situated in the second half of the twentieth century. However, by referring to the family history of the main character and the life of other, older characters, the reader is informed about the troubled history of the first half of the twentieth century for the German-speaking population of Belgium as well.

In all three novels we can detect a struggle on the part of the main character with the notion of borders. The concept of borders is indeed complex and difficult to define, as there exist different kinds of borders: 'When borders are crossed and, as a result, spaces are opened, this does not only happen in a geographical sense. It can occur quite literally [...] or in a figurative way' (Horst, 2009).[6] In this way, it is possible to distinguish between geographical, regional and political borders, language borders, cultural borders and mental borders. In this chapter I focus on politico-geographical and mental borders. Although both the socio-historical context and the general topic of the three novels are the same, the way in which they deal with the issue of borders is different. We can actually notice two different trends in the novels by Wintgens and Anderer, while Derwahl's novel can be seen as presenting an intermediary position, a kind of bridge between these two trends.

With respect to the notion of borders, the focus in *Wege aus Sümpfen* by Leo Wintgens is mainly geographical and political. The novel depicts the life of the protagonist, Peter, in his hometown on the Belgian-German border. He is the youngest of six children in a family marked by the difficult past of the region. The landscape plays such a major role in this novel that it could also be considered as a second protagonist. In both a literal and a metaphorical way, the land of the region is a moor, in which the inhabitants can easily get stuck. The image of the moor refers largely to the difficult relationship the characters maintain with the past, which they would rather forget than deal with. As a consequence they remain stuck in that past. This situation is also represented by the notion of the border, which in this novel is closely connected to the situation of the borderland.

The importance that the novel attributes to the border situation on the one hand and to geographical aspects and nature on the other is emphasised in the title: *Wege aus Sümpfen. Roman einer Grenzlandschaft*. However, we can also find the presence of mental borders. This is the case when the protagonist Peter mentions about school that: 'A wall had been erected in his innermost being, he knocked, scratched at it, but he could not break through' (p.411).[7] Through such sentences we discover not only that politico-geographical borders exist, but that borders are also internalized by individuals and exist in the minds of the characters. In this case the protagonist feels locked up, which means that he feels the presence of a border that is neither visible nor necessarily perceived as such by other characters. It is an inner mental border of the protagonist. Moreover, the novel shows that political and national borders can generate mental borders: 'There [at Garnstock], a line had been drawn across Limburg at the time of the Congress of Vienna, a line that also separated the people. However, this strict borderline did not affect all souls' (p.494).[8] So, by establishing a political border across the province of Limburg in 1815, a mental border was generated in the mind of some, but not all inhabitants.

This shift of signification of the notion of borders from a politico-geographical to a mental context follows the general evolution of the idea of belonging and *Heimat*. The German concept of *Heimat*, a complex and historically changing notion of geographical, cultural, social and affective belonging and identity, is difficult to translate into other languages and thus points to the fact that the actual identification processes denoted by this concept might indeed be specifically complex as well. However, here is not the place to dive into the particularities of German cultural history and history of mind: suffice it to mention that while from the early nineteenth century onward *Heimat* had a very strong geographical connotation, its social dimension came to be more and more recognized in the second half of the twentieth century. Later on, the time-related aspect of *Heimat* also gained more attention, implying the recognition that the longing for *Heimat* can also be a longing for a moment in the past, such as childhood, or even for an imagined moment in the future (Peinbauer, 2007, p.5). So *Heimat* can be primarily a geographical place, defined by geographical borders, a more social entity or even chiefly a mental construction, defined by social or mental borders respectively. However, one dimension does not exclude the others. *Heimat* is now mostly considered as a notion that has to be actively constructed and achieved by the individual in a process that is influenced by political, social and medial aspects (Daum, 2007, p.3).

In the analysed German-language Belgian novels, we can see that the characters encounter difficulties in identifying themselves with their geographical *Heimat*:

> [Peter] didn't know anymore, if he really belonged there or not ... Maybe he was the boy from the south, whom someone had dragged away, the prisoner of the Corsair, who only longed for one thing, that was to move to his true *Heimat*?[9]
>
> (p.134)

Peter doubts whether the place he lives in is really his true *Heimat*. He feels as if he has been 'dragged away' from his true *Heimat*, even if he cannot define where it is located. For Peter, as for the characters in the novels by Derwahl and Anderer, the geographical borders do not seem to offer a plausible framework for identification and belonging. Nevertheless the geographical borders still play a major role in Leo Wintgens's novel.

The novel *Unterwegs zu Melusine* by Hannes Anderer deals with the notion of borders in a very different way. The protagonist of this novel, Jean-Pierre, mainly experiences problems in dealing with the mentality of the population of the region he lives in, which he considers short-sighted. He longs to escape this small region in which he feels constantly confronted with borders. Since his youth he has wanted to cross these borders to get to know other cultures and he tries to fulfil this desire by undertaking cycling trips to other Belgian and German cities. This longing to cross borders could be seen as an appeal of the protagonist towards his region and its inhabitants to open up their borders, instead of closing themselves off within various kinds of borders, politico-geographical but above all mental, that are founded in historical and cultural-linguistic particularities. The novel, which is clearly influenced by existential philosophy, is full of reflections on the life path the protagonist wants to follow. In *Unterwegs zu Melusine* the notion of mental borders is thus much more present than that of geographical borders. In contrast to the currency of the word 'border' in Wintgens's novel, Anderer frequently refers to the notion of 'freedom': 'I need my free space, even if only symbolically' (Anderer, 2006, p.365).[10] We are dealing here with a protagonist who tries to escape from the oppressive environment he is living in. Jean-Pierre indeed suffers from the strong influence of the Catholic Church on society and on daily life (school, family and, not least, sexual and erotic life) and he experiences the region he lives in as a border region, in which he feels captured. On his path through life and towards greater freedom, the protagonist finds support in the existential reflections of Jean-Paul Sartre and the eighteenth-century writer Denis Diderot. He manages to achieve part of the freedom he longs for when he leaves school and, later, when he leaves home to attend university. However, in spite of the regional and mental boundary crossing, the protagonist remains caught in his past:

> Before my departure to Leuven, I found myself in a curious state of mind. On the one hand, I was again and again taken over by a delirium of freedom. I would at last take life in my own hands and shape it myself. On the other hand, a sense of hate, aversion, and bitterness took hold of me as I looked back on my past. My previous life seemed a waste of time to me. I could only see my past as a time of mental pain, denigration, domestication, in which everything individual and peculiar had to be suppressed and covered.[11]
>
> (pp.375–376)

In all three novels, the past is described as mostly being suppressed or denied and not properly dealt with. The characters remain stuck in this past and are not able to move on, i.e. they do not achieve a successful *Vergangenheitsbewältigung*; they are unable to reconcile with the identity of their *Heimat* and, as a result, with their own identity. The difficulty of identity construction is also demonstrated in the quote above: the protagonist experienced his youth as a period of his life 'in which everything that was one's own and everything that was special had to be suppressed and covered' (p.376). Through the novel we understand that the oppressive presence of the Catholic Church at school and at home played a major role in this. This feeling of being locked up can be seen as emblematic for the uneasy feeling the protagonist experiences by living in this border region and from which he tries to escape. However, he will need to be able to deal with his past before he can entirely liberate himself from the mental borders imposed on him, and thus, a successful identity construction will be possible.

A similar problematization of identity construction characterizes Freddy Derwahl's novel *Bosch in Belgien*. The main theme of this novel is the search by the protagonist for his *Heimat* and the question of where it should be situated. A geographical approach does not seem to provide an answer to this question. The loss of a sense of *Heimat* causes identity problems for the protagonist – problems that are reflected in the surname of the protagonist, 'Bosch'. 'Bosch' is an insulting nickname that the French-speaking population used for Germans from the First World War until long after the Second World War. Albert Bosch in the novel feels rejected by his country and by the people living in it. In the course of the novel it becomes clear to the protagonist that *Heimat* and its borders do not necessarily have to be understood in a geographical sense, but can also take on a more symbolic signification. Through other characters, he becomes aware of different concepts of borders and *Heimat*. In particular, Jeanne, a Brussels woman who sells chips at fun fairs, offers him a different perspective on the notion of *Heimat*. She feels at home everywhere in her country: whether in Flanders, in Wallonia, in Brussels or in the German-speaking part of Belgium. Albert falls in love with her, but as long as he does not learn to appreciate his *Heimat*, he will not be able to gain Jeanne's love. He then decides to move to Brussels, the capital and thus the symbol of his country, but also the city he despised for so long, and he starts to study the history, geography and local politics of Brussels, as well as the German expats who lived in the capital during the First World War. Thanks to this study, he is able to see the troubling history resulting from the effects of the Second World War in his country in a different light. Whereas the Second World War, which is still fresh in mind, caused material and mental destruction, the German expats living in Brussels during the First World War also enabled cultural input and exchange. It is as if the cultural context of the First World War gives Albert the possibility of overcoming the purely negative image he has kept of his home region due to the Second World War. At the end of the novel he is able to gain Jeanne's love. From the moment he realizes that one can only feel at home where one loves, he no longer searches for the local *Heimat*.[12] Love, therefore, literally replaces a geographical *Heimat* for Albert. The protagonist is able to acquire a sense of belonging by developing a more emotional and affective understanding of *Heimat*. The novel *Bosch in Belgien* is thus about the personal journey of the main character, which leads him from a mainly geographical and political concept of *Heimat*, which does not seem to be satisfying, to a more affective mental understanding. Albert learns on his journey that *Heimat* does not necessarily rely on geographical borders:

> No land, but still *Heimat*, that was actually the extract of what he had always looked for in his complicated East-Belgian domicile. That this was to be found 'anywhere and nowhere' and was furthermore timeless, matched with his feeling of longing after all the rejections, disappointments from this side and the other side of the narrow borders. If he would succeed to capture a tip of this morningland in the occidental fog, it would be for him an exterritorial, impregnable *Heimat*.[13]
>
> (p.171)

Albert frees the concept of *Heimat* from space- and time-related constrictions: *Heimat* can be found 'anywhere and nowhere' – an attitude he seems to identify here as an oriental alternative to the occidental fog.

Comparing the three novels, the novel *Bosch in Belgien* can be seen as occupying an in-between-position, building a bridge between a geographical and political notion of the concept of borders, which is very present in *Wege aus Sümpfen*, and the mental and existential notion of the border that can be found in *Unterwegs zu Melusine*.

Towards a land without borders

In this chapter I dealt with the notion of the border as a topic in the recent literature stemming from German-speaking Belgian authors. By concentrating on works by Leo Wintgens, Hannes Anderer and Freddy Derwahl I demonstrated that borders, while being a central theme in all three examples, are presented very differently by the individual authors. In *Wege aus Sümpfen. Roman einer Grenzlandschaft. Teil I* Leo Wintgens attributes great importance to geographical and political borders, while in *Unterwegs zu Melusine* Hannes Anderer primarily focuses on mental borders, and in *Bosch in Belgien* Freddy Derwahl depicts the gradual development from one understanding to the other. In the three novels the concepts and the border dimensions primarily dealt with vary: whereas in *Wege aus Sümpfen* 'border' itself is central and with it the notion of geographical borders, *Unterwegs zu Melusine* is about 'freedom' and the mental dimension of borders, and in *Bosch in Belgien* '*Heimat*' is the keyword and change the central issue. However, all three authors, each in their own way, seem to stress that it is exactly the fact of living in a borderland and reflecting on geographical and national borders that enables their protagonists to overcome a narrow politico-geographical notion of borders. Remember the 'native who always *feels* the borders better' from the poem (*l'indigène qui ressent toujours mieux les frontières*). Understanding borders as mental constructs enables reflection not only about the present situation, but also about the past, which was very turbulent in the region. The characters learn to deal with this past by being confronted with different kinds of borders, which thus become an integral part of the cultural memory of this region (Beck, 2010, p.135). Their ability to deal in a satisfactory way with the past is a necessary step in reconciling with the *Heimat*, which seems to be primordial for their identity-construction. While the three protagonists are more or less successful in their personal struggle – Jean-Pierre in *Unterwegs zu Melusine* clearly did not fully succeed in liberating himself from borders – the authors are unanimous about the fundamental necessity of struggling against these borders.

The characters of the German-language Belgian novels have to cross the borders to move on and open perspectives for the future – not only space-related borders, but also time-related borders, cultural borders and language borders. As Arvi Sepp mentions about Leo Wintgens's trilingual poem 'land without borders', 'the openness and multilingualism of the German-speaking border-space in Belgium [are] highlighted' (Sepp, 2010, p.23).[14] The message of the poem is clear, even though it contradicts the title of the novel *Roman einer Grenzlandschaft*: the borders in this 'borderland' have disappeared, not just the geographical and political borders, but also the mental and linguistic ones.

Notes

1 Leo Wintgens was born in 1938 in Hergenrath, which is now part of the East Belgian municipality of Kelmis. He obtained a PhD in Philosophy and Philology from the University of Liège and became Professor of English at the Haute École Paul-Henri Spaak in Brussels. He has published several literary and scientific works and is active on various committees, above all for East Belgian causes (Wintgens, 2001, p.504).

2 Title: land without / borders land without / borders land without / borders
First strophe: "Land / without borders" / a stroke / word for tourists / who briefly / admire / the region / from outside
Second strophe: "Land / without borders" / barriers / there are none - / remain / but the old impositions / that the native / who feels always / more the borders / has to cross / like anywhere and anytime
Third strophe: land / without / borders a STRIKE / only on / the old hollow kettledrum / that accompanies some fractious / historically fed / prejudices / in counterpoint //
[All translations of citations are mine.]

3 Freddy Derwahl was born in 1946 in Eupen. He studied literature and sociology in Leuven, Aachen and Paris. For years, he was active as a journalist, but since 2007 he has exclusively focused on writing literature (Derwahl, 2012). Hannes Anderer is the pseudonym of the author Jean Firges, who was born in 1934 in Sankt Vith. He studied German and Roman Philology in Leuven, Heidelberg, Freiburg, Genoa and Cologne. From 1970 until 1999 he was Professor of French Language and Literature at the Pedagogical College of Ludwigsburg, Germany (Strasser, 2007, p.12). He died in 2014.

4 For more detailed information on this topic, see Minke, 2010.

5 Mental borders are receiving more and more interest in social science research on borders and bordering processes; see, among others, Paasi and Prokkola's (2008) study on the maintenance of mental borders in the Finnish-Swedish border area.

6 "Wenn Grenzen überschritten und damit Räume eröffnet werden, geschieht das nicht nur im geografischen Sinn. Es kann ganz wörtlich geschehen […] oder auch im übertragenen Sinn" (Horst, 2009).

7 "Eine Wand war in seinem [Peters] innersten Wesen errichtet worden, er klopfte, kratzte an ihr, aber durchbrechen konnte er sie nicht" (Wintgens, 2001, p.411).

8 "Dort [am Garnstock] war zur Zeit des Wiener Kongresses ein Strich quer durch Limburg gezogen worden, der auch die Menschen trennte. Aber der strikte Grenzstrich ging nicht durch alle Seelen" (Wintgens, p.494).

9 "[Peter] wusste nicht mehr, ob er wirklich dazugehörte oder nicht... War er vielleicht der Junge aus dem Süden, den jemand verschleppt hatte, der Gefangene des Korsaren,

der sich nur nach einem sehnte, wieder in seine richtige Heimat zu ziehen?" (Wintgens, 2001, p.134)

10 "Ich brauche meine Freiräume, wenn auch nur symbolisch" (Anderer, 2006, p.365).

11 "Ich befand mich in der Zeit vor meinem Aufbruch nach Löwen in einer sonderbaren Verfassung. Auf der einen Seite wurde ich immer wieder von einem Freiheitstaumel erfasst. Ich würde endlich mein Leben selbst in die Hand nehmen und es selbst gestalten. Auf der anderen Seite empfand ich beim Rückblick auf meine Vergangenheit ein Gefühl von Hass, Widerwillen, Verbitterung. Mein bisheriges Leben kam mir vor als verlorene Zeit. Meine Vergangenheit sah ich nur negativ als eine Zeit der psychischen Qualen, der Verunglimpfung, der Domestizierung, in der alles Eigene und Besondere unterdrückt und zugedeckt werden musste" (pp.375–376).

12 "Zuhause war man nur, wo man liebte" (Derwahl, 2006, p.334).

13 "Kein Land und doch Heimat, das war eigentlich das Konzentrat dessen, was er in seinem verwickelten ostbelgischen Standort immer gesucht hatte. Dass dies "überall und nirgends" zu finden und obendrein zeitlos sei, entsprach nach all den Abweisungen, Enttäuschungen diesseits und jenseits kleinlicher Grenzen seiner Sehnsucht. Sollte es ihm gelingen, im abendländischen Nebel einen Zipfel dieses Morgenlandes zu erwischen, würde es ihm eine exterritoriale, uneinnehmbare Heimat sein" (p.171).

14 "In Leo Wintgens' dreisprachigem Gedicht "land ohne grenzen" [werden] die Offenheit und Mehrsprachigkeit des deutschsprachigen Grenzraumes in Belgien hervorgehoben" (Sepp, 2010, p.23).

Bibliography

Anderer, H., 2006. *Unterwegs zu Melusine*. Annweiler am Trifels: Sonnenberg Verlag.

Assmann, A., 2013. *Das neue Unbehagen an der Erinnerungskultur. Eine Intervention.* München: C. H. Beck Verlag.

Beck, P., 2010. Vergangenheitsbewältigung und Identitätssuche in der Literatur Ostbelgiens. *Belgien im Fokus – Geschichte, Sprachen, Kulturen, Band III: Die Deutschsprachige Gemeinschaft*, pp.113–138.

Daum, E., 2007. Heimat machen! Über Verbindungen von Ort und Selbst. *Heimatpflege in Westfalen*, 2, pp.1–10.

Derwahl, F., 2006. *Bosch in Belgien*. Eupen: Grenz-Echo Verlag.

Derwahl, F., 2012. *Die fliehende Frau*. Graz: Leykam.

Fickers, A., 2004. Gedächtnisopfer. Erinnern und Vergessen in der Vergangenheitspolitik der deutschsprachigen Belgier im 20. Jahrhundert. *Zeitenblicke*, 3, 9 June. Available online at: www.zeitenblicke.de/2004/01/fickers [Accessed 19 June 2015].

Horst, C., 2009. Raum- und Körperbilder in der Migrationsliteratur. Available online at: www.migration-boell.de/web/integration/47_2024.asp [Accessed 19 June 2015].

Minke, A., 2010. Grenzland seit Menschengedenken. In A. Begenat-Neuschäfer, ed., 2010. *Belgien im Fokus – Geschichte, Sprachen, Kulturen, Band III: Die Deutschsprachige Gemeinschaft*. Frankfurt am Main: Peter Lang. pp.3–26.

Paasi, A. and Prokkola, E. K., 2008. Territorial dynamics, cross-border work and everyday life in the Finnish-Swedish border area. *Space and Polity*, 12(1), pp.13–29.

Peinbauer, R., 2007. *Zurück in die Anti-Heimat. Das Heimkehrmotiv in ausgewählten Werken von Gerhard Fritsch, Franz Innerhofer, Robert Menasse und Peter Zimmermann*. Wien: Universität Wien.

Sepp, A., 2010. Grenzübergänge. Transkulturalität und belgische Identität in der aktuellen deutschsprachigen Literatur in Belgien. *Aussiger Beiträge: Germanistische Schriftenreihe aus Forschung und Lehre*, 4, pp.13–26.

Strasser, A., ed., 2007. *Mit leichtem Gepäck. Eine Anthologie ostbelgischer Gegenwartsliteratur*. Sankt Vith: edition KRAUTGARTEN.

Wintgens, L., 2001. *Wege aus Sümpfen. Roman einer Grenzlandschaft. Teil I: Eine Jugend auf dem Königshof*. Aachen: Helios.

7 The impact of commuting on close relations

Case study of Estonian men in Finland

Keiu Telve

The number of commuters between Estonia and Finland has drastically increased during the last ten years, and the mobile workforce has become integrated into both societies. The need for manual labourers in neighbouring countries, a recession in Estonia, a noticeable salary gap and better working conditions in Finland have motivated many Estonians, especially men, to consider the untraditional life-pattern of living and working in different countries. The 80-kilometre distance between Estonia and Finland, affordable ferry tickets and active sea-traffic together encourage weekly and monthly commuting between the two countries. In consequence, commuters are choosing the higher wages and better working conditions in Finland while at the same time holding on to their networks in Estonia, which means their families do not have to move and they do not have to fully integrate into a different society.

Commuting has become an important issue in public debate in recent years (Woolfson, 2007; Lukjanov, 2012; Kartau, 2013; Kalberg, 2013; McBride and Bergen 2014; Nerb, et al., 2009). It is often pointed out that the mobile lifestyle is accompanied by many problems: the distancing from one's social networks; the struggles inside the family; and an imbalance in the household workload (meaning in most cases an increase of work for the wife). My empirical research on commuting between Estonia and Finland indicates that this negative impact on family life and intimate relationships is only one facet of the experiences and perceptions of commuting. However, the positive effects have received only a little attention. The question of how these effects, most importantly the increased income, can resolve long-lasting problems in the family has not been much under discussion. In addition, the evolution of family relations, which in some cases is described as a growth of affection, has received little attention in the context of migration research.

This article is based on a qualitative case study examining the commuting practices of Estonian men in Finland. One of my main research questions was how international commuting affects close relations, bonds with family, other relatives and friends, and in this chapter I would like to give an overview of the main results. I will begin by briefly outlining the migration patterns between Estonia and Finland. I will then describe the fieldwork process and methodology, before discussing some of the research already available on the topic. In the second half of the chapter I will present the empirical findings and discuss them, noting the impact of commuting on close relations from the perspective of the men who are travelling between two countries weekly or monthly.

Work-migration between Estonia and Finland

In my study I concentrated on people who maintain residence in Estonia but who work in Finland and stay there during weekdays or for longer working periods. I will use the term 'commuter' to reference these international commuters. The distance between Estonia and Finland, the cultural similarities between the two countries and the easy journey back and forth makes commuting here similar to cases elsewhere in Europe where people are commuting inside a country. However, cross-border mobility creates a transnational sphere, where people are, through economic, social and political connections, tied to both countries (Basch, Glick Schiller and Szanton-Blanc, 2000, p.7).

Emigration from Estonia to the old European Union member states increased significantly in the 2000s compared with the first decade of Estonia's re-independence. Labour migration also increased. The cross-border work and migration of Estonian residents has been strongly influenced by two events: the country's EU accession in 2004; and the global economic crisis which started in 2008. Within around two years of the crisis, more than 106,000 people in Estonia lost their jobs, and in 2010 the unemployment rate reached an all-time high. It has been pointed out that Estonia is one of the major countries of origin of commuting workers in Europe: there are 15.8 cross-border commuters per thousand inhabitants in Estonia (MKW Wirtschaftsforschung and Empirica, 2009). In 2013, 14 per cent of the Estonian population aged over 15 years had worked or were currently working abroad, while the EU average was 9 per cent (European Commission, 2013). Finland is the main destination for Estonian migrants: 65 per cent (16,200 people) of all the migrants choose this country (Tarum, 2014, p.5). Another 5.9 per cent of Estonians (51,600 people) have thought about, or even made preparations for, working abroad. Migration and transnational commuting are more common among people from the rural and peripheral areas than among those from Tallinn and other cities (Anniste, 2014, p.20; Krusell, 2013). There are more female than male emigrants (women accounted for 54 per cent of Estonian emigrants), however, men are dominant (86 per cent) among those who work abroad but reside in Estonia (Tarum, 2014, p.4). The average age among commuters is 30 to 45 years, so it can be concluded that most of them are in a relationship or married and they have one or more kids. The majority of Estonian emigrants are employed in the construction sector, and blue-collar and skilled labour types of employment are most prevalent. According to Mõtsmees (2012) nearly half (42 per cent) of the men had to work jobs they were overqualified for or had to move downwards on the career ladder.

However, mere statistics on labour migration cannot give us a full picture of the people who work abroad. The specifics of labour migration inside the European Union are hard to track. Many people work in Estonian companies, but they do subcontracting work in Finland. Some people work in Finland without a written contract, and this illegal labour force is not visible in any statistics. Some of the work abroad is irregular and short-term, and people do not officially register this work with tax administrations or other official bodies. Given all of these possibilities, it is very likely that the actual number of Estonian commuters and work migrants is much higher than the official data suggest.

Fieldwork context and methods

My fieldwork was conducted among male Estonian blue-collar and skilled workers. Most of the subjects were in the construction sector or had previously been connected with it. Their blue-collar activity makes them part of the largest commuter group between Estonia and Finland. This group has received significant attention from social research using quantitative approaches (cf. Kalev and Jakobson, 2013; Tarum, 2014; Krusell, 2013). In this chapter I am more interested in the emic perspective, i.e. the commuters' understandings and view of their way of life.

My fieldwork lasted almost a year and consisted of internet data collection, open-ended interviews, observation and an online survey. Since the men were at first rather reluctant to talk about their personal relations, I started with online data collecting in social networks. Since December 2013 I have been part of the Facebook groups 'Estonians in Finland', 'FinEst – Estonians in Finland' and 'Estonian Entrepreneurs in Finland'. These Facebook groups can be seen as social and instrumental networks for Estonians who are living in Finland or who are regularly commuting between the two countries. The online environment facilitates contact with people who are or have been in the same situation, and who discuss the problems they face and solutions. Close relations were a regular topic in these groups, from how to spend time with their children who are visiting their parents in Finland during the holiday to which phone company to use to get the best call rates to Estonia to be able to maintain active connections with a wife and children. Many discussions centred on the daily lives of Estonians in Finland and consisted of debates about the patterns of life that would ensure the best experience for the whole family. One of the most popular posts during my fieldwork period, measured in 'Likes' and comments, was: 'But let's be honest, how many men are putting their wedding ring into the pocket while travelling to Finland?' There are tens of posts daily, making this medium ideal for studying the Estonians' integration into Finnish society as well as their problems and needs.

In order to further address personal issues in the research, I opted for a combined methods approach, conjoining an online survey with subsequent in-depth open-ended interviews with a small subsample of participants. Sampling selection was based on circulating the survey through social media, especially Facebook networks. The most useful were the aforementioned groups 'Estonians in Finland' (6,169 members), 'FinEst – Estonians in Finland' (20,125 members) and 'Estonian Entrepreneurs in Finland' (3,600 members), which made it possible to contact many people simultaneously. All together there were 149 responses to the questionnaire, of whom 40 per cent were men. The age proportions mirrored the work migration statistics – most people were between 26 and 45 years old. The questionnaire consisted of twenty-five questions and the answers were anonymous. The most important and valuable parts of the survey were the open-ended questions, which, in some cases, were answered in-depth, providing informative qualitative data. It can be assumed that the online survey was distanced enough and people did not worry about their privacy and were therefore more willing to give information about sensitive topics, such as close personal relations.

The most detailed information about close relations came, however, from open-ended interviews that I held during nine months from the winter of 2013 through to the summer of 2014. The online survey supported the construction of the questionnaire by facilitating the selection of key topics. Interviews consisted of six thematic fields, of which one concentrated on close personal relations, although family and friends also came under discussion in the context of other questions. During the fieldwork I had contact with thirteen men and their families and friends. The sample included skilled and blue-collar workers who had worked in Finland for as short as a few months and as long as over ten years. Most of the participants had a long-term partner or were married (seven out of ten). In one case my informant had recently moved to Finland with his partner, and in another cross-border commuting was the reason why the partnership had ended. Eight participants had children and in six cases the children were minors. All the participants maintained daily or at least weekly contact with their families, often through telephone calls, and they also chatted and spoke with their families through internet communications (Facebook, Skype, Messenger etc.). The occasions when participants were able to spend weekends or even longer periods with their families in Estonia or Finland were highly valued.

Finally the analysis of newspaper articles and internet forums was important in my fieldwork process. I analysed the articles that were connected with commuting, especially commuting between Estonia and Finland, and mapped the main topics and narratives. Specific attention was given to the forums relating to the newspaper articles, where people could give their comments. Those included a mixture of personal experiences relating to the topic of the article and more general opinions, the latter providing an approximate insight into public opinions on and perceptions of commuting and commuters.

Research on commuting and close relationships

The impact on close relationships of commuting and working abroad is a topic of some discussion in academia in Estonia and also in other Baltic and European countries. Among others, Charles Woolfson (2007) and Anna Broka (2009) have studied close relationships in Baltic countries; Anna White (2010) has worked on Poland; and commuting related to Russian industry (particularly the Russian oil industry) has received attention through Edward Holland and Eldar Eldarov's (2012) and Gertrude Eilmsteiner-Saxinger's (2011) research. There has also been previous research on Estonia: Age Viira has studied the commuter's family life from the point of view of partners who are left behind (Viira, 2010), and Margit

Kartau (2013) and Helen Pärna, Karmen Lai and Taimi Tulva (2008) have given attention to the children's perspective, asking how working abroad affects children in the family.

There is a general lack of consensus when it comes to the impact of commuting on personal relationships. Half of the authors mentioned above stress that international commuting always has a negative impact on close relations with friends and family. This point of view is defended, for example, by Woolfson, who stresses that the periodic absence of one family member is accompanied by diminishing intimacy among people, resulting in families growing apart. A keyword in his work is 'mushroom orphans', which refers to children left behind in Latvia in the custody of grandparents and friends by their parents who go to work in Ireland's mushroom farms (Woolfson, 2007). A similar pattern is described by White, who uses the term 'Euro-orphans' to characterise the modern Polish family model where parents work in Central-European countries (White, 2010). Kartau has found from the previous quantitative research on Estonian families that in most cases parents who start to work abroad leave their children in the home country, and contact with them diminishes over time. Being left behind can be connected to later feelings of loneliness, stress, emotional incapability and discipline problems (Kartau, 2013, quoted in Tarum, 2014). Based on qualitative research, White has also noted psychological and behavioural problems among Polish children whose parents work in another country (White, 2010).

Another frequently mentioned problem is the negative impact of commuting on intimate relationships and marriages. In the Estonian case, Viira has found that when the man begins working abroad, the women's workload, as far as housework and children are concerned, increases, resulting in stress. Homesickness and loneliness are also problems for both partners (Viira, 2010). One of the main problems in relationships is a decrease in affection and greater emotional distancing that can end in separation or divorce. Constant moving between two countries is also sometimes connected with parallel partnerships or even parallel families. For example, Andrea Rossi has shown that the stress of being alone can lead to engagement in a relationship with a new partner from the working country, which might have devastating effects on the family back home (Rossi, 2008).

In addition to these critical perspectives on the effects of work migration, more positive research results are also to be found. An often-mentioned aspect is that increased economic stability decreases the tensions resulting from financial strains, and the extra income can improve the well-being of the whole family in the long term. For example, Eilmsteiner-Saxinger (2011) shows that commuter-workers are more active consumers of free time facilities, spending more money on entertainment and everyday needs. Viira has also stressed positive effects, such as the way in which economic stability results in fewer misunderstandings and fights about how money should be used (Viira, 2010, p.43). Commuters also save for their children's education and put reserves aside to secure income in old age, both of which would be impossible for these families without the extra income that comes from working abroad (Viira, 2010, p.33). Another interesting result seen in empirical studies is that people value their time with their family more when they are working abroad than when both partners live continuously with their family in the same country. The time that is spent together is given extra value, and to make it even more special family members try to plan their mutual time actively, whether it is going on holiday together, taking mini-breaks, visiting friends or other such activities (Viira, 2010, p.12). Some people find that living periodically apart from close ones can even increase the quality of being together as a family while at the same time bringing all the possibilities connected with extra income during the holidays. Holland and Eldarov (2012) also stress that seasonal migration to an area where salaries are higher could help to raise the family's quality of life in the long term because the extra income makes it possible to buy a house or start a family business.

From the previous research it can be seen that the impact of cross-border commuting on close relationships is connected with the length of the period that one partner is staying abroad, the frequency of visits back home, as well as the circumstances of family decision-making. For example, when both partners agree that the separation is for a future goal, perhaps to earn money for home renovations, it is emotionally easier to bear (Viira, 2010, p.49). Maintaining constant contact with close ones during the commuting and working period makes being apart easier, which is easy and affordable thanks to recently developed technological possibilities. Commuters further find that everyday communication can help to maintain their relationships and can decrease the feeling of geographical distance (Viira, 2010, p.49).

The impact of commuting on close relations: The case of Estonian commuters in Finland

Given the results of previous research, it should be clear that the impact of commuting cannot be understood as only positive or negative. The situation is much more complex, justifying a qualitative approach to include in-depth research on commuters' perspectives.

The male part of the relationship – i.e., in most of the cases, the commuter himself – is largely neglected in academic research on the familial effects of commuting. In this chapter I therefore concentrate on how men see the effect of their constant moving on close relations, and what they do to decrease the potential problems. First, I describe emotional distancing and problems that are connected with it. Second, I examine the valuation of family time. Third, I consider how work in Finland and the extra money earned by men raises the quality of life of the whole family and can thus compensate for some of the discomforts of commuting. I conclude the chapter with future perspectives: how people who are at the moment actively moving between two countries imagine their lives in the near future.

Both the aforementioned studies and the majority of Estonian media reports propagate the idea that working in Finland has mostly negative outcomes for commuters' family lives. Negative narratives are also present among my participants. The overall viewpoint that is repeated over and over again is that mobility effects include the deterioration of relations between partners, parents and children or friends. Distancing from acquaintances and not being present most of the time can also diminish social capital (Bourdieu, 2008). When the network of friends and acquaintances fades away, people lose opportunities to draw on social connections productively because the previous relations do not help them to find a job, say, and commuters might find themselves left without the possibility of giving or receiving favours. One of the participants explains this well:[1]

> It goes without saying that relations with those close ones who stay in Estonia are going to get worse or even end. Marriages and relationships break up, children and parents become strangers, friends meet infrequently and relations vanish.
>
> (Online questionnaire answer, 2013)

The quote brings out that the negative impact of commuting is visible among different types of close relations. Based on the results of my fieldwork, however, it can be argued that relationships within the nuclear family, i.e. with wife and children, are seen as most vulnerable. These relations need everyday contact, and staying apart is emotionally exhausting for all. The tensions inside the family may be the reason why people either stop commuting or split up:

> When one member of the family goes to work abroad, how long does it last, what do you think? One goes and another one stays behind, how long does she wait? A year? Two? And at the end they are apart. I see it every day, I have seen a lot of times in ten years, what happens with families when Estonian men are coming to Finland without their wives.
>
> (Online questionnaire answer, 2013)

As mentioned above, double families are a favourite topic in social media discussions on commuting. In direct encounters with commuters, however, they were rarely mentioned. If at all, the topic was brought up in an impersonal way of speaking. For example, Mart says:

> I am sure that you have heard how one wife is here, in Finland, and another family with kids and wife back in Estonia. I have heard it too, many times. You never know how long this kind of long-term relationship lasts. Different people suit different things.
>
> (Mart, 45 years old, interview)

Some of the participants are sceptical about the possibility of this kind of life. Hard work and long working hours make double lives particularly difficult. Kristjan and Silver's discussion touches on accounts of double families:

> Kristjan: I have heard that some guys have one wife in Estonia, another in Finland.
> Silver: It is only a myth. Probably it is not true.

> Kristjan: Some definitely have. But I believe it is rather an exception than a rule.
> Silver: I have given it a lot of thought and figured how it is possible to have a wife also in Finland, how can I afford it. I work three hundred hours per month and I spend my free time in Estonia. It would be impossible.
>
> (Silver and Kristjan, both 32 years old, interview)

As can be seen, these men doubt that this way of life is possible. It must be kept in mind that because of the scandalous nature of the topic, exceptional cases might quickly be overemphasised and generalised. The 'double families' have become a (stereo)typical characteristic of men who are working in Finland, but the results of the fieldwork imply that it is not a widespread phenomenon and commuters rarely mention concrete cases.

The most important negative outcome of commuting is the lack of time for family life. This way of being together and being apart works for some families but not for others. Marko puts it like this:

> It depends. For some people it is okay, but it is still stressful. While being abroad your home life suffers. At the end there is no 'home life'. You are more like a visitor.
>
> (Marko, 45 years old, interview)

While it cannot be argued that negative impact of commuting on close relations does not exist, possible positive effects of work migration do also exist and have not received much attention in Estonia so far. Improvements in quality of life, extra income and, in some sectors, fixed holidays are some of the reasons why people who commute may be in a better position to take care of their families. The questionnaire results show that the majority of the respondents find that their relations stayed the same (56 per cent) or even improved (24 per cent) after they started working abroad. Only 20 per cent of people stated that their relations with close ones had deteriorated.

Many of the informants find that being abroad has made them appreciate their relationships more. They talk with their partner on a daily basis – mostly about everyday issues – and this kind of communication is an important part of maintaining the emotional relationship. Furthermore, men who commute have found that mobility has made them more understanding of their partner. Most of the time, being together but being apart needs more planning, calculating and flexibility to meet the partner's needs. Some of the interviewees have even mentioned that their marriages have improved. For example, Margus describes it as follows:

> Commuting and living in different countries makes you appreciate your close ones. Before, a quarrel could come out of nothing. Some small thing could be the starting point for a fight. Now it is not so. You value every minute that you are at home.
>
> (Margus, 48 years old interview)

Laptops and smart phones are commonplace and help commuters and their families stay in touch around the clock. At the same time, new environments bring out the need to communicate with family and friends. This can be one of the reasons why people sometimes speak with their loved ones even more frequently when abroad. This is especially the case in the context of the wider family: people Skype with their parents, siblings or even with more distant relatives. Video calling can create new kinds of intimacy that are not the same as being together, but can partly counteract the effects of geographical distance. Margus says:

> We are video Skyping every day, before going to sleep. There are phones. It is so easy nowadays. Earlier it was much more difficult. There were phones, but calling was expensive. Now it is much cheaper in Finland, international calls to Estonia, to Norway and to everywhere.
>
> (Margus, 48 years old, interview)

Margus has been commuting for more than a decade and can appreciate the present situation in comparison with ten years ago, when keeping in touch was much more difficult.

The participants also talk about how, for them, planning a holiday can help them endure the work and separation. Planning events with the family can create the feeling of togetherness. Due to the lack of time spent together, people value these moments more and are ready to put significant effort into the planning and preparations, so that all the family members enjoy them and the free days are spent as efficiently as possible. For example, Kristjan explains:

> Our holiday plans are made weeks before. That way there is not this kind of just hanging out and doing nothing. So that time would not slip away. [...] We love to travel, my wife and I, we visit friends. This kind of usual stuff.
>
> (Mart, 45 years old, interview)

Fathers often state that working abroad is something they do for their children and family. It is most of the time framed as altruistic behaviour, the economic stability of the family being the most commonly mentioned reason for commuting work. The majority of men who are working in Finland felt the effects of the economic crisis of 2008–2009 in Estonia, so social and economic security are most valued. The present income helps them to support their family with everything needed, including holidays, a car, property and, very importantly, higher education for their children.

> I am working there because of them [my family]. It is not for me. For my sake I could sit here all day long. It would be no problem to work in Estonia.
>
> (Silver, 32 years old, interview)

In addition to the economic improvements, various forms of social security also play an important role. Finland's welfare system helps residents in case of unemployment: people are able to maintain the same or almost the same level of income through government support for an extended period after losing their job. It protects workers in the long term and thus also enables them to make plans, take out a bank loan, enlarge their family etc. Some interviewees even claim that these long-term security aspects are more important to them than monthly wages.

Economic security and extra income make various entertaining and holiday activities reachable. The heads of the family take pride in knowing that they can afford these amenities, and they also point out explicitly that stress about money inside the family has disappeared.

> Everyday problems have disappeared. How to survive from payday to payday. You do not have to borrow 20 euros from your friend to buy food. Now there is a sauna in my apartment, car loan is paid within two years instead of five. Life has improved.
>
> (Margus, 48 years old, interview)

Extra income can be spent on a home, a car or paying back loans. As previously mentioned, another important sphere of spending is travelling. Income helps workers to go on vacation with the whole family, including young children. It can be said that as a result of the commuting, families spend time together more actively:

> You can go everywhere. To a theme park with children, to visit some place. The whole family together. During the autumn we went to Turkey with the family, to make the summer a little bit longer. It was the first foreign trip with the four of us together. This year we want to go again.
>
> (Tõnu, 38 years old, interview)

As seen from previous quotes, the income from Finland can improve the quality of life for the whole family. Another aspect that has not yet been mentioned is the economic support provided to grown children and elderly parents. Men are mostly very aware of the financial situation of their parents, who often have to live on small pensions, and they try to help them whenever they can.

> Of course I am supporting my parents. I know these pensions. I could not do it on an Estonian wage.
>
> (Kristjan, 45 years old, interview)

The same way of thinking can be observed with regard to grown children who earn small salaries or study. Commuter fathers not only support them in their immediate needs but also try to secure their long-term wellbeing. Commuters invest in buying apartments, accruing some additional income for retirement. Most of the participants say that their families have accepted that they are working abroad and that they support them in all ways possible. From the men's point of view, their wives and children know that working abroad is a decision that has been made for the wellbeing of the whole family and this helps them to accept the hardships of geographical separation more easily.

I want to stress that the positive and negative aspects of commuting cannot be seen separately but as tightly connected with each other. This means that the keyword 'distance' has a double meaning in the context of cross-border commuting. Even though the absence of one family member *per se* is hard to bear, distance is also connected with heightened appreciation of the missing member or the time spent together. Additionally, being away helps men to see their close relationships more clearly and sometimes leads them to consider ways to improve them. At the same time, however, they are not present long enough to change things:

> When you are away, you can see your life from a different perspective. [...] It may even motivate you to change things, if you see them from the distance. But another question is, how much you can actually do, but at least you aware of them.
>
> (Tõnu, 38 years old, interview)

In both cases we can see how the distance is seen at the same time as an advantage and a disadvantage for family life.

Even though most of the commuters indicated that their relationships have not got worse, they admitted that they would prefer to be in a situation where all the family members were in the same country. Many of them say that the majority of their commuting friends have, after some period of travelling back and forth, chosen Finland as their permanent place of residence. Eventually most of the migrants are able to access the social and economic infrastructure in the receiving country, allowing them to secure a more permanent life for themselves and their family members, who might follow them (Castles and Miller, 2009, p.29). Commuting between two countries costs both money and time, and a lot of couples eventually reach the decision to stay put. Because of the higher salaries and social benefits, they often prefer to live in Finland. At the same time, however, there are a lot of people who find that commuting has become their way of life. They do not want to move to Finland because their home is in Estonia or their partner's life choices are holding her there. It is also difficult to move to another country with school-aged children. For those families, the best solution seems to be commuting. Although many studies argue that commuting is a non-permanent way of life (Chiswick, 2000; Massey et al., 1993), in the Finland-Estonia case it seems that rather the opposite is true. When these commuters choose a country, at the moment they opt for Finland.

Conclusion

This chapter has discussed the impact of Estonian-Finnish work migration on close personal relationships. Because of current Estonian migration patterns, it is a very relevant topic for both Estonian and Finnish society. The analysis of Estonian media and internet forums shows that the overall opinion about commuter families is rather negative. Numerous examples indicate that the wider public tends to connect commuters with stereotypes that presume that a mobile lifestyle is always negative for the family and the wider social field: couples split up, parents and children lose their connection and friends become strangers. Based on the results of qualitative research among male commuters I was able to demonstrate that the phenomenon is much more heterogeneous. Although the participants indicated that there are negative aspects, such as missing their loved ones, and that the mobile way of life is not 'family suitable', they also mentioned a number of positive effects of working abroad. Commuters point out that they appreciate family time and their close ones more. Economic stability and securing a better quality of life for the whole family in the long term were the primary benefits discussed. It was also argued that by enabling men to solve financial problems, commuting also enhances familial harmony and stability. A large proportion of commuting can be argued to be motivated by altruism – men earn money abroad first because of, and for, their close ones. However, even those who say that their relationships have not changed during the commuting period would prefer the situation where all the family lives in the same country. In the present Estonia-Finland context, those commuter families who ultimately turn to a more sedentary and close family life very often do this by moving to Finland. In these cases, the period of commuting can thus be seen as a transitory phase in the process of migration.

The research results indicate that the effects of commuting on family life and intimate personal relationships are complex and do not allow unequivocal conclusions. In the presented study, research has been focused on the experiences and perceptions of the commuters themselves – more precisely, on male commuters, a group that has clearly been neglected by research on commuting and family life so far. It goes without saying that the presented work is just a first step and that further qualitative research on these particular family situations is needed that takes into account the various actors involved. Furthermore, it is important to widen the research perspective by comparing the Estonian-Finnish case with other examples of (cross-border) commuting.

Note

1 The online questionnaires were completed anonymously; quotes from in-depth interviews are supplemented with a pseudonym and the age of the individual participant.

Bibliography

Anniste, K., 2014. East-West Migration in Europe: The Case of Estonia after regaining Independence. *Dissertationes geographicae Universitatis Tartuensis*, 57, Tartu: University of Tartu Press.

Basch, L., Glick Schiller, N. and Szanton-Blanc, C., 2000. *Nation Unbound. Transnational Projects, Postcolonial Predicaments and Dererritorialized Nation-States*. London, New York: Routledge.

Bourdieu, P., 2008. The Forms of Capital. In N. W. Biggart, ed., 2008. *Readings in Economic Sociology*. Malden, MA: Blackwell. pp.280–291.

Broka, A., 2009. *Impacts of Labour Migration on Children Left-behind in Latvia: The Role of the State*. Oslo: Oslo University College.

Castles, S. and Miller, M. J., 2009. *The Age of Migration.* New York: Palgrave Macmillan.

Chiswick, B. R., 2000. Are Immigrants Favourably Self-Selected? An Economic Analysis. In C. B. Brettell and J. F. Hollifield, eds., 2000. *Migration Theory: Talking Across Disciplines*. New York, London: Routledge, pp.61–76.

Eilmsteiner-Saxinger, G., 2011. 'We feed the nation': Benefits and challenges of simultaneous use of resident and long-distance commuting labour in Russia's Northern Hydrocarbon Industry. *Journal of Contemporary Issues in Business and Government*, 17(1), pp.53–67.

European Commission Internal Market, 2013. *Special Eurobarometer* 398. Available online at: http://ec.europa.eu/public_opinion/archives/ebs/ebs_398_en.pdf [Accessed 4 March 2015].

Holland, E. C. and Eldarov, E. M., 2012. "Going away on foot" once again: The revival of temporary labour migration from Russia's Dagestan. *Central Asian Survey*, 31(4), pp.379–393.

Kalberg, S., 2013. Seelik lehvis ringi kui mees Soomes töötas. *Eesti Ekspress,* 22 August. Available online at: http://ekspress.delfi.ee/news/paevauudised/seelik-lehvis-ringi-kui-mees-soomes-tootas.d?id=66618341 [Accessed 15 July 2014].

Kalev, L. and Jakobson, M., 2013. Hargmaisus Eesti-Soome ruumis. *Acta Politica Estica*, 4, pp.95–113.

Kartau, M., 2013. *Lapse igapäevaelu peale mõlema vanema välismaale tööle siirdumist: lapse perspektiiv.* Bachelor's thesis, University of Tartu, Faculty of Social Sciences and Education, Institute of Social Studies.

Keely, C. B., 2000. Demography and International Migration. In C. B. Bartell and J. Hollifield, eds., 2000. *Migration Theory. Talking Across Disciplines.* London, New York: Routledge. pp.43–60.

Krusell, S., 2013. Eesti elanike töötamine välismaal, In: Statistikaamtet, 2013. *Pilte rahvaloendusest. Census Snapshot.* Tallinn: Statistikaamet. pp.129–146.

Lukjanov, G., 2014. Miks Soomes töötavad Eesti mehed oma naisi petavad. *Postimees, Naine 24.* Available online at: http://naine24.postimees.ee/717066/miks-soomes-tootavad-eesti-mehed-oma-naisi-petavad [Accessed 15 July 2014].

Massey, D. S., Arongo, J., Hugo, G., Kouaouci, A., Pellegrino, A. and Taylor, J. E., 1993. Theories of International Migration: A Review and Appraisal. *Population and Development Review*, 19(3), pp.431–466.

McBride, M. and Bergen K. M., 2014. Voices of Women in Commuter Marriages: A Site of Discursive Struggle. *Journal of Social and Personal Relationships*, 17(February), pp.1–19.

MKW Wirtschaftsforschung and Empirica, 2009. *Scientific Report on the Mobility of Cross-border Workers within the EU-27/EEA/EFTA countries*. Available online at: www.euregio-heltal.org/wp-content/uploads/2010/09/EU-Commission-Cross-border-workers-in-Europe-2009.pdf [Accessed 15 May 2015].

Mõtsmees, P., 2002. *Välismaal töötamise mõju edasisele töökarjäärile.* Presentation in Estonian National Bank. Available online at: www.mtk.ut.ee/sites/default/files/mtk/dokumendid/Pille%20M%C3%B5tsmees.pdf [Accessed 16 July 2014].

Nerb, G. F., Hitzelsberge, F., Woidich, A., Pommer, S., Hemmer, S. and Heczko, P., 2009. *Scientific Report on the Mobility of Cross-Border Workers within the EU. Final report.* Brussels: European Commission DG Employment and Social Affairs Brussels.

Pärna, H., Lai, K. and Tulva, T., 2008. Vanemate töötamine välismaal- kuidas see mõjutab lapse elu? *Sotsiaaltöö*, 6, pp.36–44.

Rossi, A., 2008. *The Impact of Migration in Developing Countries*. Presentation in Youth Migration Conference. Available online at: http://crcw.princeton.edu/migration/files/Rossi.pdf [Accessed 15 April 2013].

Tarum, H., 2014. Eesti tööealise elanikkonna väljarändepotentsiaal aastal 2013. *Sotsiaalministeeriumi toimetised*. 2. Tallinn: Sotsiaalministeerium sotsiaalpoliitika info ja analüüsi osakond, pp.1–34. Available online at: www.sm.ee/sites/default/files/content-editors/Ministeerium_kontaktid/Valjaanded/eesti_valjarandepotentsiaal_2013_2.pdf [Accessed 20 July 2014].

Telve, K., 2015. *Kvalitatiivne lähenemine pendeltööle Soomes töötavate Eesti meeste näitel*. Master's thesis, University of Tartu, Institute of Arts and Culture Studies.

Viira, A., 2010. *Kui mehed on kaugel tööl ehk piiriüleste pendeltöötajate pered Eestis*. Bachelor's thesis, Tallinn University, Institute of International and Social Studies.

White, A., 2010. Young People and Migration from Contemporary Poland. *Journal of Youth Studies*, 13, pp.565–580.

Woolfson, C., 2007. Labour Standards and Migration in the New Europe: Post-Communist Legacies and Perspectives. *European Journal of Industrial Relations*, 13(2), pp.199–218.

Part III
Living across the border

8 Cross-border residential mobility of people working in Luxembourg

Developments and impacts

Birte Nienaber and Isabelle Pigeron-Piroth

Introduction

The integration of the European Union and the freedom of movement inside the Schengen area have had deep impacts on mobility in border regions. One region that has been especially influenced is the Greater Region Saar-Lor-Lux. This region comprises Luxembourg, the German federal states of Rhineland-Palatinate and Saarland, Lorraine in France and Wallonia in Belgium. According to the Observatoire Interrégional du Marché de l'emploi, 213,400 people inside this geographical area live and work in two different countries. They constitute one quarter of the total number of cross-border commuters inside the European Union (EU27) (OIE, 2014).

Luxembourg is the economic driving force of the Greater Region. The attraction of the Luxembourgish labour market not only leads to high commuter rates but also to a steady demographic increase, and therefore also to an increase in real estate prices near the Luxembourgish border. These important cross-border flows create interdependencies between territories. Indeed, the Luxembourgish economy relies on the cross-border workforce that constitutes about 44 per cent of the salaried population. Half of these cross-border commuters come from France, a quarter from Belgium and the last quarter from Germany. Luxembourg provides employment to many people living on the other side of the border. In the French region of Lorraine, one in ten employees has a job outside France. In northern Lorraine, near the border, this proportion is significantly higher and can be as much as 80 per cent for some small villages in the immediate vicinity of the border (Belkacem and Pigeron-Piroth, 2015). This phenomenon is structural and has many consequences for territories on both sides of the border with regard to economic development, spatial planning and so on. This specificity of the cross-border situation needs to be taken into account when exploring these territories (Belkacem and Pigeron-Piroth, 2012).

The cross-border residential mobility we are studying in this chapter is linked to Luxembourg and its economic role within the Greater Region. Analysing residential mobility is complex, because there are many explanatory factors and interactions between them. Factors linked to housing (e.g. size, type, geographical location) can cause the mobility, but so can familial ones (marriage, divorce, the birth of children) as well as the characteristics of employment (type of activity, sector of activity) and its localization (place of work, distance from home to work). Moreover, our case study is specific because of its cross-border focus. The borders create differentials (especially in housing prices and the availability of housing) that can explain the moves to the other side of the national border. Furthermore, residential moves and daily commutes from home to work are interrelated, and moves are often the result of striking a balance between residential mobility and commuting (Pigeron-Piroth, 2008). Each factor influences the others. On the one hand, good connections (by public transport, by motorways, etc.) favour longer distance daily journeys from home to work and extend the residential area of the commuters. When anchoring to a place is really important, long commutes can be an alternative to residential moves. Travel time often tends to be taken as the relevant factor rather than geographic distance. On the other hand, the burden of the commutes can cause a residential move closer to the place of work. Our chapter analyses the developments of the cross-border residential mobility of people working in Luxembourg and living in the neighbouring countries, from quantitative as well as qualitative perspectives.

State of the art

Nowadays, there is widespread consensus in the social sciences and humanities that borders should not be understood as given lines on a map but as socially constructed realities. In this context, bordering, including de-bordering and re-bordering, has become an important research concept relating to processes along borders that are mainly influenced or heightened by globalization. Having been established in the mid-1990s, this approach is now very well developed in border studies (e.g. Laine, 2012; Newman, 2006; Paasi, 2011; Sendhardt, 2013; Van Houtum, Kramsch and Zierhofer, 2005). Bordering separates and brings together. Borders allow certain expressions of identity and memory to exist while preventing others. They are open to being contested at the level of the state and in everyday life (Paasi, 2011, p.62). Thus, borders can disappear and other (new) borders appear, and these 'processes of de-bordering and re-bordering are not exclusionary, but occur simultaneously' (Laine, 2012, p.51).

This conceptual reorientation implies a change in research focus 'shifting from border territories to border behaviour' (Terlouw, 2012, p.353). The notion of 'border surfers' describes one such behaviour that is constituent of de-bordering processes. Border surfers are 'very specific groups of transnationals [that] base their way of life on profiting from the different opportunities on both sides of the border through regular border-crossing' (Terlouw, 2012, p.354). As an ideal case study Terlouw (2012) presents cross-border residential mobility which is a research topic that has gained in importance, especially since the 2000s. It deals particularly with people who move to a neighbouring country while still working in their home country. Christian Wille (2012) calls these people 'atypical cross-border commuters' since, unlike typical cross-border commuters, they live abroad and commute to their country of origin for work. Several cross-border areas in

Europe have been analysed with respect to residential mobility – for example, the Belgian-Dutch (e.g. van Houtum and Gielis, 2006), the Danish-Swedish (e.g. Jagodic, 2012), the French-Swiss (e.g. Rérat et al., 2011), the German-Dutch (e.g. Strüver, 2003; van Houtum and Gielis, 2006; Gielis and van Houtum, 2012), the German-Luxembourgish (e.g. Carpentier, 2010; Nienaber and Kriszan, 2013; Wille, Schnuer and Boesen, 2014; Nienaber and Frys, 2015; Roos et al., 2015), the German-Polish (e.g. Jonda, 2009; Balogh, 2013; Nienaber and Kriszan, 2013), the Italian-Slovenian (e.g. Jagodic, 2012), and the Slovakian-Hungarian borders (e.g. Hardi, 2008). This list is not complete and can provide only a small insight into a research topic that is increasing in importance at the same time as the phenomenon of cross-border residential mobility increases in Europe, where the internal borders are losing their barrier effects.

From the booming country of Luxembourg, cross-border mobility takes place into all three neighbouring countries: Belgium, France and Germany – especially into the rural areas. Brosius and Carpentier (2010) analysed settlements in all these three countries and came to the conclusion that municipalities like the German town of Perl, where many Luxembourgers already live, will gain even more Luxembourgers in what they regard as a continuing trend – a so-called chain migration. The main reasons for moving across the border into neighbouring countries are financial (real estate prices constitute the main reason for 50 per cent of people), family related (family formation etc.) or cultural (e.g. school system, language skills) (Carpentier, 2010). Housing prices are indeed higher in Luxembourg than in the surrounding border regions of Germany, Belgium and France. Becoming a homeowner is the second most cited main motivation (11 per cent). For 10 per cent, growth in the size of the household (through marriage, etc.) was the most important factor, and for 7 per cent separation was of primary importance, showing that familial reasons are also important in the decision to move. A qualitative research study in 2009/2010 demonstrated that while cheaper living and real estate costs are the main reasons for moving to the neighbouring German municipality of Perl, Luxembourgers who moved to the other side of the national border are still deeply rooted in their home country and in their work in Luxembourg. This attitude works against integration into the new residential town (Frys and Nienaber, 2011).

Methodology

In the following, we will give a more detailed account of the recent development in cross-border mobility from Luxembourg into the neighbouring borderlands. In a first step, large quantitative datasets are analysed for the purpose of stocktaking, as well as to study the main evolutions and characteristics of cross-border residential mobility. We used data files from the social security system of Luxembourg[1] to identify residential cross-border moves. All the people working in Luxembourg (including the cross-border commuters) who are registered in the Luxembourgish social security system are taken into account. This means that the data presented in this chapter do not count non-working individuals (for example, children or retired persons). Since the individuals have moved to or from Luxembourg while continuing to work in Luxembourg, it can be reasonably assumed that they relocated within commuting distance. Therefore, the data are not concerned with residents who relocated to or from Luxembourg and also changed their country of employment. The variables are socio-demographic (age, nationality etc.), geographic (place of residence) and economic (sector of activity etc.).

To analyse the development of the phenomenon of residential mobility around Luxembourg, different longitudinal data files for three-year periods (from 1995 to 2013) were constructed. The first period is 1995–1998 and involves 178,450 people working in Luxembourg in March 1995 *and* in March 1998. The last period is 2010–2013 involving 301,450 people. This comparison of six periods allows us to measure the evolution of residential moves for people working in Luxembourg over the last twenty years, in keeping with a previous study we conducted for the period 1995–2005 (Pigeron-Piroth, 2008).

We then focused on another file containing 195,000 people working in Luxembourg in March 2003 *and* in March 2013. For each of them, their place of residence in 2003 as well as their characteristics can be compared with the same information ten years later. People living in Luxembourg in 2003 and living abroad in 2013 were thus analysed in detail.

This quantitative information is supplemented by qualitative guided interviews, to provide a deeper, more individualized view of factors such as reasons for migration. Forty qualitative semi-structured interviews were carried out with people of Luxembourgish nationality living in Perl between 2009 and 2012. These interviews were transcribed and then analysed using content analysis methods.

Development of cross-border mobilities between Luxembourg and its neighbouring regions

From 1995 to 2013, six three-year periods were used to study the development of residential moves for people working in Luxembourg. The residential mobilities were identified by comparing the country of residence in year Y and in year Y+3.

The graph in Figure 8.1 shows the number of residential moves from Luxembourg to Germany, France and Belgium, and the reverse movement from abroad to Luxembourg. In the two first periods, the majority of cross-border residential moves for people working in Luxembourg was oriented towards Luxembourg. In recent periods, however, the trend has changed in the direction of the neighbouring countries. Indeed, from period to period, the numbers of people working in Luxembourg and leaving the country to settle abroad have been increasing. During the period 2007–2010, concerned by the economic crisis, a maximum number of 3,500 employed people (including 1,100 Luxembourgers) left Luxembourg. This is more than three times the number in the first period. However, the last studied period (2010–2013) is characterized by two specificities: a decrease in departures from Luxembourg (for the first time) and an increase in people settling in Luxembourg. The new settlements abroad decreased in

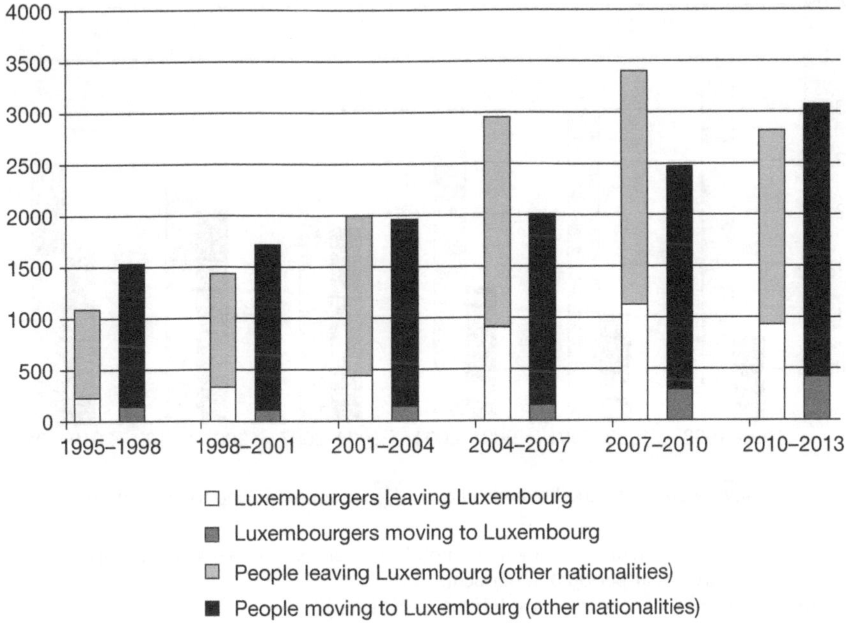

Figure 8.1 Development of residential mobilities for people working in Luxembourg (three-year periods).
Source: University of Luxembourg, IGSS/CCSS (Data).

comparison to the previous periods, especially for the moves to Germany for Luxembourgers and Portuguese, the latter constituting the most important group of foreign residents in Luxembourg. Does this last period indicate a new trend after the huge increase in the two previous periods? The number of people moving to Luxembourg is increasing from period to period, and became higher than the number of outmovers in the years 2010–2013. Part of this recent movement (15 per cent) is due to Luxembourgers who lived abroad and then came back to Luxembourg (this aspect will be developed in the following section).

To better understand this growing trend, it is also relevant to express the developments as a percentage of the overall population – that is, of the 'potential movers' (Figure 8.2). The period 2007–2010 appears with the maximum percentage of settlements abroad (2 per cent of the resident population working in Luxembourg), before showing a decrease in the following period. Thus during the last period studied (2010–2013), 1.6 per cent of the people living and working in Luxembourg left to settle abroad, while 2.5 per cent of the cross-border commuters moved to Luxembourg.

Figure 8.2 also suggests an inverse relationship between the evolution of the percentage of people leaving and moving to Luxembourg, implying that at least some circumstances that reduce the rate of people moving to Luxembourg might also be relevant to those leaving Luxembourg. Structural factors would be good explanatory starting points to explore this inverse relationship, since, for example,

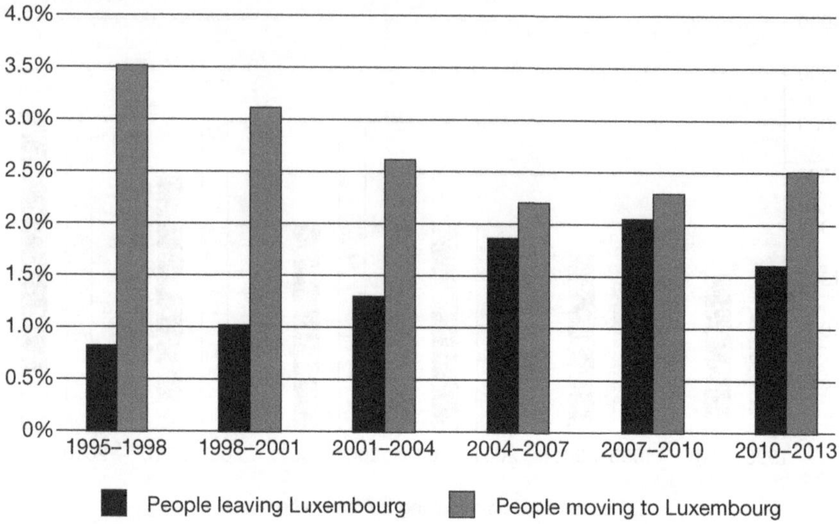

Figure 8.2 Development of residential mobilities for people working in Luxembourg (three-year periods) in per cent.
Source: University of Luxembourg, IGSS/CCSS (Data).

rising real-estate prices in Luxembourg and better cross-border infrastructure would explain both a decrease in people moving from the borderlands to Luxembourg *and* an increase in the number of people leaving Luxembourg. This suggests that, in addition to traditional explanatory factors (like housing, familial factors, etc.), these residential moves are likely to be linked to the distances and travel times between home and work and the housing prices in the Greater Region as a whole. The next part of this chapter will centre on residential mobility from Luxembourg to the neighbouring countries, with a view to gaining a better insight into the characteristics of these new cross-border commuters.

People leaving Luxembourg and becoming cross-border commuters (2003–2013)

This section focuses on a period of ten years. People who worked in Luxembourg in March 2003 *and* in March 2013 were identified, as well as the characteristics of those people within this group who moved from Luxembourg to one of the neighbouring countries between these two dates. Out of 195,195 persons working in Luxembourg in 2003 and ten years later, we identified 5,150 people who left Luxembourg but continued to work there.[2] That means that 4 per cent of the people living and working in Luxembourg in 2003 moved to settle abroad during the subsequent ten-year period and still lived abroad in 2013. This is much more than during the period 1995–2005, in which 2 per cent of working residents moved abroad (Pigeron-Piroth, 2008). The residential moves are almost evenly split between the three bordering countries: most are to Germany (35 per cent), then

France (34 per cent) and finally Belgium (31 per cent). By contrast, during the period 1995–2005 most of the departures from Luxembourg were to France (43 per cent), while 26 per cent chose Germany. The attractiveness of Germany increased greatly up to 2010. The population leaving Luxembourg is not homogeneous, and the choice of the new country of residence varies according to nationality (Table 8.1).

Luxembourgers leaving Luxembourg

The majority of people moving abroad between 2003 and 2013 were Luxembourgers (35 per cent).[3] This residential mobility involves a change of status for them: they become cross-border commuters to their own country of origin. Some authors consider them as 'atypical' cross-border commuters because their commute across the borders is not common and is linked more to the real-estate situation than to the labour market (Wille, 2012). The number of Luxembourgers living abroad and working in Luxembourg has increased greatly in recent years (Figure 8.3).[4] However, they represent less than 3 per cent of the population of cross-border commuters, whereas at the French-German border, the German employees living in France (Moselle-Est) and working in Germany (Saarland) constitute one-third of the cross-border commuter flows (OIE, 2014).

The Luxembourgish candidates for residential cross-border mobility are of great interest for the real estate agencies, which engage in strategic advertisements of properties on the other side of the border (Figure 8.4). The poster advertising

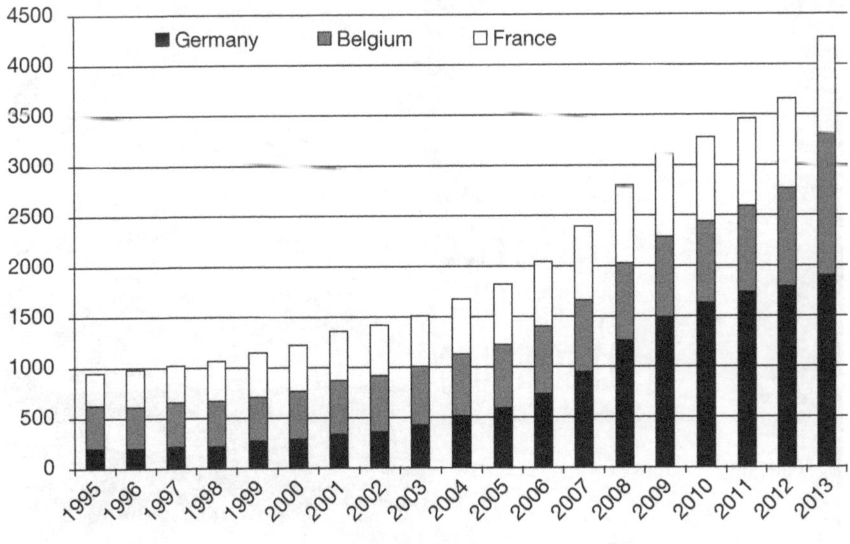

Figure 8.3 Luxembourgers living in Germany, France and Belgium and working in Luxembourg.
Source: University of Luxembourg, IGSS/CCSS (Data).

new buildings in France (left hand side of Figure 8.4) was situated inside the main station of Luxembourg City, through which a large number of people pass each day. The main message was 'live close to Luxembourg'. Indeed, the agency SODEVAM proposed various new construction projects in France, near the Luxembourgish and German borders and close to the main motorways. The second advertisement (right-hand side) offers apartments in the German municipality of Perl, promoted, built and sold by a Luxembourgish company.

Germany is indeed the most frequently chosen country of residence for Luxembourgers leaving Luxembourg (more than two thirds) (Table 8.1). This can be explained by the similarity of the language and the proximity to Luxembourgers already settled there (who attract new Luxembourgers), among other factors. Before the move, most of the Luxembourgers leaving Luxembourg were living in the capital or in the south of the country. After the move, in 2013, 12 per cent of those who moved live in Perl, a small German town close to France and Luxembourg, and a further 11 per cent live in four other German towns near the border (Mettlach, Freudenburg, Merzig and Wincheringen). A special focus on Luxembourgers living in Perl is included in the following section.

Most of the Luxembourgers leaving Luxembourg work in public administration, or transport and communication. The Luxembourgish employees who are the most likely to move abroad are those from the transport sector (mainly the state railways), the construction sector and industry. By contrast, civil servants, farmers, the self-employed and those in liberal professions are less involved in such moves.

Figure 8.4 Advertisements for real-estate from both sides of the border.
Source: The authors.

Table 8.1 Country of residence in 2013 for people leaving Luxembourg, according to nationality

Nationality	Country of residence in 2013 (in %)			
	Belgium	*Germany*	*France*	*Total*
Belgian	*89.1*	3.4	7.4	100.0
German	4.2	*91.0*	4.8	100.0
French	13.3	4.9	*81.8*	100.0
Italian	34.2	18.1	*47.7*	100.0
Luxembourgish	14.2	*67.6*	18.2	100.0
Portuguese	24.1	12.3	*63.6*	100.0
Others	21.4	*46.4*	32.2	100.0
Total	*30.8*	*35.2*	*33.9*	*100.0*

Base: People working in Luxembourg in March 2003 and 2013, living in Luxembourg in 2003 and abroad in 2013.
Source: University of Luxembourg, IGSS/CCSS (Data).

French, Belgians and Germans

After the Luxembourgers, the main nationalities of people leaving Luxembourg are French (21 per cent) and Belgian (21 per cent). German people constitute just 6.5 per cent of the population leaving Luxembourg to settle abroad. These people lived in Luxembourg in 2003 (they either previously relocated to Luxembourg or they were born in Luxembourg) and left Luxembourg between 2003 and 2013. The new country of residence is closely related to their nationality: 82 per cent of the French employees leaving Luxembourg decided to settle in France, 89 per cent of the Belgians in Belgium, and 91 per cent of the Germans in Germany (Table 8.1). The settlement of the French employees in France is less geographically concentrated in one municipal area close to the Luxembourg border. They mainly choose bigger municipalities that are readily accessible by train or motorways (such as the municipalities of Thionville and Hettange-Grande), rather than being directly on the borders. In Belgium, there is a concentration of cross-border migrants in the city of Arlon, close to the Luxembourg border (19.5 per cent of the people who moved to Belgium lived here in 2013). The Germans leaving Luxembourg settle in the city of Trier (15 per cent) rather than the municipality of Perl (8 per cent), which was, however, mainly chosen by Luxembourgers.

Finance, services and trade are the main sectors of activity of French and Belgian people leaving Luxembourg. German people leaving Luxembourg mainly work in the financial sector as well as in the trade and industry sectors. Salaries, working hours, place of work and accessibility to work are some of the factors linked to the activity in Luxembourg that might explain the move abroad. The liberal and self-employed professions are underrepresented in the population leaving Luxembourg, regardless of their nationality.

Portuguese

As already mentioned, Portuguese residents are the largest group of foreign nationals living in Luxembourg; they constituted 16 per cent of the population in 2015 (STATEC). Their numbers in Luxembourg began increasing in the 1970s when many Portuguese were fleeing the dictatorship in Portugal and were looking for work and better living conditions.

The Portuguese are the fourth main nationality involved in cross-border residential moves from Luxembourg to the neighbouring countries in the period under review (2003–2013): 9 per cent of the people leaving Luxembourg are Portuguese. They generally choose French-speaking countries: 64 per cent of them lived in France in 2013 and 24 per cent in Belgium. Most of them settled in French municipalities directly on the border. They mainly work in the construction sector, followed by the service sector and trade.

The move back to Luxembourg

We will conclude our overview by looking at the move back to Luxembourg for Luxembourgers who live abroad (and work in Luxembourg), tracing its development from period to period. Taking into account the six periods of three years (from 1995–1998 to 2010–2013), the number of Luxembourgers moving to Luxembourg is increasing, as is the percentage they constitute, especially for the two most recent periods. During the first period (1995–1998), about 150 Luxembourgers working in Luxembourg and living abroad in 1995 moved to Luxembourg. They constituted 9.5 per cent of the population moving to Luxembourg. In the last period (2010–2013), more than 400 Luxembourgers moved to Luxembourg from abroad and constituted approximately 14 per cent of the total population moving to Luxembourg. The two last periods we studied show an increase in the number (and the percentage) of Luxembourgers moving to Luxembourg. Among the population of Luxembourgers living abroad in 2010 and working in Luxembourg, 14.5 per cent of them moved to Luxembourg between 2010 and 2013.

Some of the characteristics of this population during this last period (2010–2013) may be pointed out. Most of these people lived in Germany in 2010 and precisely nine per cent lived in the town of Perl before their move to Luxembourg. Their main places of residence after the move to Luxembourg in 2013 are the city of Luxembourg and the main municipalities of the south (30 per cent live in the five most populated cities of Luxembourg). In 2013, the average age of this population was 39 years,[5] and a clear majority were men (61 per cent). Most of them work in public administration and in the social sector (39 per cent), and in fact they are overrepresented in these sectors in comparison to the Luxembourgers who lived abroad in 2010 and 2013.

Having given a general insight into the recent development of cross-border residential mobility and the complex constitution of the group that moves from Luxembourg to a neighbouring border region, we now turn to the motivations for these moves. We will do this by turning to one specific locality, the German municipality of Perl, and we will see that in the cross-border area under study, real estate prices do not constitute the sole explanation for residential mobility. Family, schooling, commutes and other factors also have to be taken into account in order to understand the cross-border moves. The town of Perl, one of the main destinations for Luxembourgers moving abroad, is an interesting case study, where the motivations and new lives of these cross-border commuters were analysed by using data stemming from qualitative interviews. The study in question did not involve the whole group of new residents from Luxembourg; rather, as mentioned above, it was explicitly focused on persons with Luxembourgish nationality.

Booming border municipalities – the example of Perl

Perl is a small town in the German-Luxembourgish-French border triangle. Between 2005 and 2007, the growth in the Luxembourgish population in this town was remarkable (between 30 per cent and 40 per cent each year) (Figure 8.5). This development can be partly explained by easier access to Luxembourg via a motorway and bridge crossing the Moselle River which was completed in 2003. After 2008 the Luxembourgish population in Perl has continued to grow, but more slowly, especially in recent years. On 1 January 2015, among the 8,920 registered inhabitants of Perl, 20.1 per cent had Luxembourgish citizenship. They made up 64.6 per cent of all foreign nationals in the municipality. The unofficial numbers might be even higher, as it is possible that a significant number of Luxembourgers do not register in Germany. They often remain registered in Luxembourg to continue to qualify for access to Luxembourg's social welfare support and public services, especially unemployment pay and schooling for their children, as many interviewees explained.

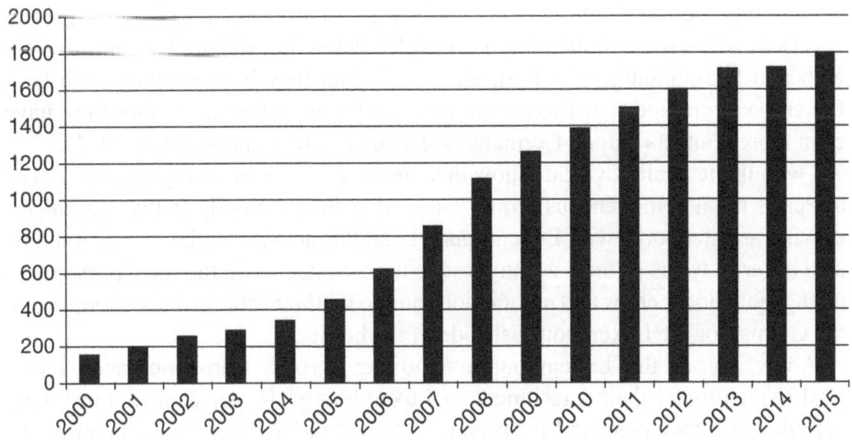

Figure 8.5 Development of Luxembourgish nationals living in Perl (2000–2015).
Source: University of Luxembourg, Municipality of Perl (Data).

Inside the municipality of Perl its fourteen villages are affected by the phenomenon of Luxembourgish migration to a greater or lesser degree, the most affected villages being Perl-Münzingen (30.4 per cent of Luxembourgers), Perl-Sinz (26.6 per cent) and Perl-Perl (25.0 per cent).[6] The age structure, family structure and employment situation of the people interviewed were heterogeneous. People with young children were among the interviewees as well as retired individuals who came to Germany to stay in a retirement home. Luxembourgers earning only the minimum wage in Luxembourg were represented, as were highly skilled professionals.

The reasons for the move across the border to Germany are diverse. During the qualitative interviews, however, real estate prices were mentioned very often and can be seen as one of the main factors, or even the main factor, in the decision-making process underlying the cross-border move.

Reasons for choosing Perl are also diverse: family reunification ('Our granddaughter lives over there' (interview 30); 'My son bought a house here in Perl and that is why I moved here' (interview 26)); the already existing large Luxembourgish community in Perl (mentioned during several interviews, e.g. interview 26); no language barrier since the dialect spoken in Perl is closely related to the Luxembourgish language (mentioned by many interviewees, e.g. interviews 17, 21 and 30); no cultural distance between Luxembourgish culture and the culture in the region in and around Perl (e.g. interview 29); the German-Luxembourgish Schengen-Lyzeum (a German-Luxembourgish secondary school created in 2007 where pupils from both Luxembourg and Germany are educated in their respective national curricula and school system or in a merged bilingual schooling system); the landscape (e.g. interview 27); and good transport connections to Luxembourg (e.g. interviews 20 and 26).

However, these numerous reasons for moving to Perl do not imply that the bonds to the interviewees' former place of residence have diminished. The results of our qualitative interviews show that those who have moved still maintain many or all of their personal contacts in Luxembourg. As for contacts in Perl, the picture is varied. While some of the interviewees stated that they did not have any contact with non-Luxembourgers in Perl, others said that they have contacts with both Luxembourgers and non-Luxembourgers, and some also stated that they have even more contacts among Germans (see also Nienaber and Kriszan, 2013).

Overall, the qualitative data show that the group of Luxembourgers who moved to Perl is not a homogeneous group. They differ in age, family status, reasons for moving and networks with Luxembourgers and/or non-Luxembourgers, and they also differ in terms of their willingness to have contact with the local population, to engage in local clubs and associations and to do their routine daily shopping on the German or the Luxembourgish side of the border.

Analysing all the Luxembourgers moving abroad, Carpentier and Gerber (2012) mentioned their attachment to Luxembourg: 46 per cent of their daily activities still take place in Luxembourg after the move. In comparison, only one fifth of the Germans leaving Luxembourg maintain activities – other than work-related ones – in Luxembourg after the move. As for the Belgians and the French who move away from Luxembourg, the numbers are somewhat higher: one quarter and one third, respectively.

Another notable aspect concerns the increasing number of people who decide to move back to Luxembourg, as shown in the quantitative analysis. An issue that might contribute to this decision is the feeling of not being welcome in the new place: 'The local people in Perl think that the Luxembourgers destroy the infrastructure here because we have more money to build more luxurious homes than they build' (interview 19). The same interviewee also gives another reason why people move back to Luxembourg: 'Yes indeed, the retirement pension is problematic. The taxes[7] are much higher here and therefore you do not have any money left later on. That is why many people sell their houses shortly before retirement and go back to Luxembourg' (interview 19).

The interviews thus illustrate that the reasons for moving back are as varied as the reasons for the initial move and thereby confirm one basic result of the detailed comparative analysis of the demographic data: namely, the insight that residential migration is a complex phenomenon that cannot be understood as a simple linear development.

Conclusion

More and more cross-border labour and residential moves are being observed in border regions. The Greater Region is one of the borderlands that is most affected by cross-border labour mobility. Residential cross-border mobility has also increased in the Greater Region. Associated with this is the move out of Luxembourg to the Belgian, French or German borderlands, especially for Luxembourgers, who become atypical cross-border commuters. While the Luxembourgish cross-border commuters constitute only 3 per cent of the total population of cross-border commuters, their increase in numbers over the last few years deserves our full attention.

This chapter examined the development of this phenomenon of cross-border residential mobility, focusing on people who left Luxembourg in the last twenty years and moved to neighbouring countries but still work in the Grand Duchy, providing some insights about this form of mobility from its very beginning. Since 2000 the number of people working in Luxembourg and leaving it to settle abroad has been increasing (especially until 2007–2010). However, the last period we studied (2010–2013) reveals a decrease in the departures from Luxembourg, which remains to be validated (or invalidated) by analysis of the subsequent periods.

Focusing on people working in Luxembourg in 2003 and ten years later, we identified some of the main characteristics of those who moved. Due to the heterogeneity of this population, these characteristics are not easy to summarize. Luxembourgers constitute the bulk of people settling abroad (35 per cent) and becoming (atypical) cross-border commuters. More than two-thirds of them chose Germany as their new country of residence, in both small towns directly at the border (like Perl and Wincheringen) and bigger ones near the border (Mettlach and Merzig). Luxembourgers constitute 20 per cent of the population of the town of Perl (8,900 inhabitants) and 30 per cent in the village of Perl-Münzigen.

The qualitative interviews reveal the difference in the real estate prices as the main reason to settle abroad, even though the motivations are frequently various and multiple, including family-related, cultural (language, schooling, etc.) and geographical (proximity, accessibility, etc.). Accessibility to Luxembourg, notably the proximity of train stations and motorways, is one of the main factors that influences the choice of the new place of residence. The interviews also show that the decision for living abroad is not necessarily definitive. Reasons for moving back can be relationships with the local population, burdens of the commutes or concerns about retirement. These are relatively new phenomena that will have to be investigated in future research. To mention just one aspect, only a few cross-border commuters are retired (in comparison to the total number of cross-border commuters), but there will be more and more in the future. Thus, in 2013, 28.4 per cent of the pensions paid by Luxembourg are given to people living in France, Germany and Belgium (i.e. to former cross-border commuters) (Social Security of Luxembourg), but this proportion is increasing from year to year.

To conclude, we would like to briefly return to our opening remarks on general conceptual developments in the realm of border studies, regarding the question of how far the rise in cross-border residential mobility can be seen as contributing to a process of debordering. More specifically, are those who participate in this process 'border surfers', people who act upon the principle of profiting 'from the different opportunities on both sides of the border' (Terlouw, 2012, p.354)? An important result of our quantitative analysis, which mainly consisted of the comparison of different three-year periods in an overall timeframe of nearly twenty years, was that it shows how difficult it is to arrive at a general assessment of the demographic developments and their impact on bordering processes. The same holds true for our qualitative study. Here we see that motivations and causal factors are diverse and intricately interrelated. This means that the individual actors, the persons or families who take up residence on the other side of the border, do not only behave as border surfers in the strict sense mentioned above – seeking opportunities on both sides of the border – but also rely on other qualities. In other words, it means that the qualities they are looking for on one side and the other are far from being constant. They depend upon, among other things, the development of the family and the changing needs of its members, but they can also be affected by what living abroad actually offers these surfers – in certain cases, the experience of not being welcome.

Notes

1 Inspection Générale de la Sécurité Sociale (IGSS) and Centre Commun de la Sécurité Sociale (CCSS).

2 The total number of residential moves in this period is in fact greater; we are studying only those who worked in Luxembourg both in 2003 and ten years later. All the people who were not working in Luxembourg in 2003, or who stopped working in Luxembourg between 2003 and 2013, as well as non-workers, are not taken into account in this analysis.

3 The nationality taken into account is the nationality of the people in 2003, before the move. Indeed, during this period of ten years, 8,000 people actually changed their nationality (for 94 per cent, this was to become Luxembourgers), as a result of easier access to Luxembourgish nationality (Loi du 23 Octobre 2008 sur la nationalité luxembourgeoise).
4 Part of the evolution between 2012 and 2013 in Belgium is due to changes in nationality: many Belgians acquired Luxembourgish nationality.
5 The average age in 2013, while the actual moves might have taken place before 2013.
6 Data from the municipality of Perl.
7 This person mentions taxes on salaries and pensions.

Bibliography

Balogh, P., 2013. Sleeping abroad but working at home: Cross-border residential mobility between transnationalism and (re)bordering. *Geografiska Annaler: Series B, Human Geography*, 95(2), pp.189–204.

Belkacem, R. and Pigeron-Piroth, I., eds., 2012. *Le travail frontalier au sein de la Grande Région Saar-Lor-Lux. Pratiques, enjeux et perspectives.* Nancy: PUN-Presses Universitaires de Lorraine.

Belkacem, R. and Pigeron-Piroth, I., 2015. Un marché de l'emploi intégré ? L'emploi frontalier et ses dimensions socioéconomiques. In C. Wille, ed., 2015. *Lebenswirklichkeiten und politische Konstruktionen in Grenzregionen. Das Beispiel der Grossregion Saar-Lor-Lux Wirtschaft-Politik-Alltag-Kultur*. Bielefeld: Transcript. pp.39–57.

Brosius, J. and Carpentier, S., 2010. Die grenzüberschreitende Wohnmobilität von in Luxemburg ansässigen Erwerbstätigen: Quantifizierung und Charakterisierung des Phänomens. In S. Carpentier, ed., 2010. *Die grenzüberschreitende Wohnmobilität zwischen Luxemburg und seinen Nachbarregionen.* Luxembourg: Saint Paul, pp.15–36.

Carpentier, S., ed., 2010. *Die grenzüberschreitende Wohnmobilität zwischen Luxemburg und seinen Nachbarregionen.* Luxembourg: Saint Paul.

Carpentier, S. and Gerber P., 2012. Structure des mobilités locales et reconfiguration des espaces de vie transfrontaliers luxembourgeois. In R. Belkacem and I. Pigeron-Piroth, eds., 2012. *Le travail frontalier au sein de la Grande Région Saar-Lor-Lux, Pratiques, enjeux et perspectives*. Nancy: PUN-Presses Universitaires de Lorraine. pp.313–332.

Chilla, T., Evrard, E. and Schulz, C., 2012. La gouvernance transfrontalière-territoire(s) et territorialité(s). In R. Belkacem and I. Pigeron-Piroth, eds., 2012. *Le travail frontalier au sein de la Grande Région Saar-Lor-Lux, Pratiques, enjeux et perspectives*. Nancy: PUN-Presses Universitaires de Lorraine. pp.349–365.

Frys, W. and Nienaber, B., 2011. Die Situation der Wohnmigranten im ländlichen Saarland. In H.-P. Hege, Y. Knapstein, R. Meng, K. Ruppenthal, A. Schmitz-Veltin and P. Zakrzewski, eds., 2011. *Schneller, öfter, weiter? Perspektiven der Raumentwicklung in der Mobilitätsgesellschaft.* Hannover: Akademie für Raumforschung und Landesplanung, pp.94–104.

Gielis, R. and van Houtum, H., 2012. Sloterdijk in the House! Dwelling in the Borderscape of Germany and the Netherlands. *Geopolitics*, 17, pp.797–817.

Hardi, T., 2008. *Transborder Movements and Relations in the Slovakian-Hungarian Border Regions*. Discussion Paper 68. Pècs: Centre for Regional Studies of Hungarian Academy of Sciences.

Jagodic, D., 2012. Living (Beyond) the Border: European Integration Processes and Cross-border Residential Mobility in the Italian-Slovenian Border Area. In D. J. Andersen, M. Klatt and M. Sandberg, eds., 2012. *The Border Multiple: The Practicing of Borders between Public Policy and Everyday Life in a Re-scaling Europe*, Farnham: Ashgate, pp.201–218.

Jonda, B., 2009. Neue Formen der Migration und des Austausches im grenznahen Bereich. *Polen-Analysen*, 49(09), pp.2–6.

Laine, J., 2012. Border Paradox: Striking a Balance between Access and Control in Asymmetrical Border Settings. *Eurasia Border Review*, 3(1), pp.51–79.

Newman, D., 2006. The lines that continue to separate us: Borders in our 'borderless' world. *Progress in Human Geography*, 30(2), pp.143–161.

Nienaber, B. and Frys, V., 2015. The rural regions of the old-industrialized Saarland – between globalization and regionalization. In J. McDonagh, B. Nienaber and M. Woods, eds., 2015. *Globalization and Europe's Rural Regions*. Farnham: Ashgate, pp.165–181.

Nienaber, B. and Kriszan, A., 2013. Entgrenzte Lebenswelten: Wohn- und Arbeitsmigration als Ausdruck transnationaler Lebensentwürfe im deutsch-luxemburgischen und deutsch-polnischen Grenzraum. *Raumforschung und Raumordnung*, 71 (3), pp.221–232.

Observatoire Interrégional du Marché de l'emploi, 2014. *Situation du Marché de l'emploi dans la Grande Région, Neuvième rapport de l'Observatoire Interrégional du Marché de l'Emploi pour le quatorzième Sommet des Exécutifs de la Grande Région.*

Pigeron-Piroth, I., 2008. La mobilité résidentielle des personnes travaillant au Luxembourg (1995–2005). *Bulletin du STATEC*, 2, pp.51–105.

Paasi, A., 2011. Borders, theory and the challenge of relational thinking. In C. Johnson, R. Jones, A. Paasi, L. Amoore, A. Mountz, M. Salter and C. Rumford, 2011. Interventions on rethinking "the border" in border studies. *Political Geography*, 30(2), pp.62–63.

Rérat, P., Moine, A., Gertsch, K. and Signoret, P., 2011. La mobilité résidentielle transfrontalière dans l'Arc jurassien franco-suisse, *Géo-Regards*, 4, pp.153–168.

Roos, U., Hermes, C. and Nienaber, B., 2015. Wohnsituation der atypischen Grenzgängerinnen und Grenzgänger an der luxemburgisch-deutschen Grenze. In B. Nienaber and U. Roos, eds., 2015. *Internationalisierung der Gesellschaft und die Auswirkung auf die Raumentwicklung – Beispiele aus Hessen, Rheinland-Pfalz und dem Saarland.* Hannover: Akademie für Raumforschung und Landesplanung, pp.83–99.

Sendhardt, B., 2013. Border Types and Bordering Processes: A Theoretical Approach to the EU/Polish-Ukrainian Border as a Multi-dimensional Phenomenon. In A. Lechevalier and J. Wielgohs, eds., 2013. *Borders and Border Regions in Europe.* Bielefeld: transcript. pp.21–43.

Strüver, A., 2003. Presenting Representations: On the Analysis of Narratives and Images Along the Dutch-German Border. In E. Berg and H. van Houtum, eds., 2003. *Routing borders between territories, discourses and practices.* Aldershot: Ashgate, pp.161–176.

Terlouw, K., 2012. Border Surfers and Euroregions: Unplanned Cross-Border Behaviour and Planned Territorial Structures of Cross-Border Governance. *Planning, Practice and Research*, 27(3), pp.351–366.

Van Houtum, H., Kramsch, O. and Zierhofer, W., 2005. *B/ordering Space.* Aldershot: Ashgate.

Van Houtum, H. and Gielis, R., 2006. Elastic migration: The case of Dutch short-distance transmigrants in Belgian and German borderlands. *Tijdschrift voor Economische en Sociale Geografie*, 97(2), pp.195–202.

Wille, C., 2012. *Grenzgänger und Räume der Grenze. Raumkonstruktionen in der Großregion Saar-Lor-Lux.* Frankfurt am Main: Peter Lang.

Wille, C., Schnuer, G. and Boesen, E., 2014. Beyond Luxembourg. Raum- und Identitätskonstruktionen im Kontext grenzüberschreitender Wohnmigration. In C. Wille, R. Reckinger, S. Kmec and M. Hesse, eds., 2014. *Räume und Identitäten in Grenzregionen. Politiken – Medien – Subjekte*. Bielefeld: Transcript, pp.333–345.

9 Dwelling in (un)familiarity

Examples from the Luxembourg–German borderland

Elisabeth Boesen and Gregor Schnuer

Introduction

The internal national borders of the European Union have become increasingly open in the last twenty years, leading, among other things, to a rise in the cross-border mobility of persons. Central to this are the Schengen Agreements and various directives and regulations[1] that enable EU citizens to cross national borders for travel, leisure, work and residence. This article looks at cross-border residential mobility, a particular form of cross-border mobility where people relocate their place of residence a short distance across a national border, often maintaining close ties to the country of origins, particularly through work and employment but also through family and social relationships. More specifically, we look at people who are moving from Luxembourg into the neighbouring borderland along the Moselle River in the German federal state of Rhineland-Palatinate. The data come from interviews with individuals, couples and families who have relocated into one of four German villages where the research was conducted. The border in question and Luxembourg's borders in general, even though somewhat exceptional for reasons elaborated below, are not unique in experiencing this cross-border residential mobility. Such mobility can be observed at several European borders that feature particular disparities, especially where an urban area is adjacent to a national periphery, with resulting marked differences in real estate prices.[2]

Explaining why people do or do not cross national borders is not an easy endeavour, however. We will draw on the concept of '(un)familiarity' developed by Spierings and van der Velde based on research on cross-border consumption (Spierings and van der Velde, 2008; 2013). By looking at data from our research along the above mentioned border, we hope to explore whether the concept of (un)familiarity proves fruitful to the analysis of a form of cross-border mobility that is very different from the border crossing practices considered in Spierings and van der Velde's work, and, if this is the case, whether the way (un)familiarity features in residential mobility might critically inform further conceptual work.

There are two reasons why the case of cross-border residential mobility, particularly in Luxembourg, can offer a valuable addition to the debate on (un)familiarity. First, relocating the home is a fundamentally different activity from border crossing for consumption purposes, since it implies a more or less permanent shift from one side of the border to the other, and since dwelling is in and of itself associated with ideas of 'familiarity'. Second, Luxembourg is unique both in the composition of its population and workforce, and in that it offers – by virtue of its small size – three nearby national borders, and therefore a threefold option to people considering residential migration. (Un)familiarity, in this case, goes along with relations and encounters that are significantly more complex and cannot be understood as structured by distinct experiences of a national dichotomy.

'(Un)familiarity' – why do and don't we cross borders?

In spite of continued efforts to promote border crossing within the Schengen Area of the EU, the rate of cross-border mobility of people across these borders is relatively low.[3] Spierings and van der Velde posit that crossing the border – other than staying at home – requires a particular motivation. They propose a specific approach to the analysis of this motivation by introducing the notion of '(un)familiarity' (Spierings and van der Velde, 2008). (Un)familiarity has since gained some prominence in studies on border crossing[4] as it appears to offer a compelling perspective to better understand and explain why the opening of national borders and increased cross-border integration does not necessarily result in the expected cross-border mobility.

Spierings and van der Velde show that mobility can be partially explained by what they call a 'bandwidth of (un)familiarity' (2008, pp.501–2). Push and pull factors for mobility and keep and repel factors for sedentariness act together and define that bandwidth, i.e. a range of proportions of these factors that promote cross-border mobility the most, whereas both the overly familiar and the too unfamiliar have a restraining effect. Hence, too much integration along a border may lead to increased cross-border immobility, a feature described as 'border paradox' by Spierings and van der Velde (2008, p.503). As a border opens further, people might cross it less.

The insight gained from the analysis of cross-border shopping is intuitively understood: a certain degree of unfamiliarity with what can be found on 'the other side' acts as a push factor, while too much familiarity might hamper mobility. The problem begins when we try to understand what exactly is meant by familiarity and unfamiliarity. The examples given by Spierings and van der Velde seem, at first, revealing. People do their shopping on the other side of the border if prices are lower, if the offer is more abundant etc. (Spierings and van de Velde, 2008, p.499). Unfamiliarity is thus associated with difference. Yet, by choosing their term, Spierings and van der Velde alert us to the fact that the relation they are interested in is not about difference as such, but something more complicated.

A tendency to synonymize or confound familiarity and unfamiliarity with similarities and differences can, however, be observed in much of the work in question, as well as in the research drawing on it. For example, Klatt writes in his work on labour commuters that the concept of (un)familiarity 'is used to explain how differences (*unfamiliarity*) encourage or discourage cross-border mobility'

(2014, p.354). These and other inadvertent equations between unfamiliarity and difference[5] seem to suggest that the specific content of (un)familiarity is difficult to grasp. Since the first presentation of the notion of (un)familiarity in 2008,[6] Spierings and van der Velde have been working on its clarification; a brief summary will, however, show that the problem of the relation between unfamiliarity and difference is far from resolved.

In the first instance, Spierings and van der Velde differentiate between 'rational' and 'emotional' differences (2008, p.501) and suggest possible factors that can evoke feelings of (un)familiarity: 'a lack of high-quality goods in places "at home"' and 'lower prices for the same goods in places on "the other side" can be a rational reason for international mobility,' whereas 'the "overfamiliarity" of places in the "home" country' and the 'expected extra pleasures from shopping in "foreign" places can be an emotional reason to cross the border' (2008, p.501). In a more recent article, the authors expand the concept by noting the dynamism and fluidity of (un)familiarity (Spierings and van der Velde, 2013). They reformulate 'rational' and 'emotional' differences into 'informational' and 'experiential' differences, thus allowing for (un)familiarity to be dynamic depending on changes in personal knowledge and experiences of national differences. Changing perceptions of national differences can modulate and shift (un)familiarity in a qualitative way – Spierings and van der Velde speak of attractive and unattractive unfamiliarity, or comfortable and uncomfortable familiarity, where repeat visits might turn an unattractive difference into an attractive one or vice versa. Furthermore, they introduce the notion of 'familiarisation' and thus a distinct level of dynamism. Repeat visits that create 'both more informational and experiential familiarity' (2013, p.12) not only bring about qualitative changes in the perception of unfamiliar things – for example, from attractive to unattractive unfamiliarity – but they result in a reversal: something perceived to be unfamiliar becomes perceived as familiar.

So far, these endeavours at clarification have not been altogether successful. (Un)familiarity remains a floating concept, denoting both (perceived) similarities/differences and the effects of similarities/differences, namely, a specific feeling in relation to what is expected or encountered at the other side of the border. In a recent article, however, Spierings and Szytniewski have enlarged their treatment of the subject by asking how 'familiarity and/or unfamiliarity with people and places come into being and develop during daily encounters in borderlands' (Szytniewski and Spierings, 2014, p.342). According to them 'feeling (un)familiar is a relational construct which develops through a dynamic interplay between several dimensions of (un)familiarity' (ibid.). The somewhat tautological character of the formula notwithstanding, the differentiation of several dimensions of unfamiliarity, as called for here, seems plausible. However, the scheme proposed – experiences, knowledge, self-assessment (in regard to knowledge) and proximity (p.342) – is in more than one respect open to criticism. To take but the last of these four dimensions, proximity: according to the authors proximity 'reflects the likelihood that some cultures, places and situations feel more familiar than others' (p.343). These cultures, places and situations, though familiar, may nevertheless be unknown. The problematic nature of the notion becomes clear when the authors give a concrete specification: 'A degree of proximity ... could be felt through similarities in surroundings and architecture, ethno-linguistic expression or cultural practices' (ibid.). With this definition the notion of difference creeps in yet again, proximity being more or less equated with similarity. This means that the concept is used in a metaphorical sense and its non-metaphorical pertinence is therefore disregarded. But, one might ask, do feelings of proximity necessarily depend on similarities? Such feelings may very well arise in situations where similarities in the above sense are largely or completely lacking. A non-borderland example could be illuminating here. German inhabitants of Berlin-Kreuzberg feel proximity to their Turkish neighbours although they have almost no interaction with them, know next to nothing about their circumstances and share very few socio-cultural similarities.[7] Here 'proximity' stems from proximity in its proper sense; that is, from the fact that the other is a stable part of one's inhabited space – one's home, as it were. The concept of proximity applies here in a more concrete sense; it is, so to speak, metaphorical to a lesser degree. A comparable relation may also be found in borderlands, the non-similar but familiar other being in this case the 'neighbour' on the other side of the border.[8] Like multi-cultural metropolises, borderlands offer a social environment that favours a habitus of indifference vis-à-vis difference (Bhatti et al., 2011).

To conclude our brief conceptual discussion, we want to mention a recent article by Dorte Jagetić Andersen in which she proposes an approach to the complicated interrelation between difference and (un)familiarity that comes close to our own understanding, in that (un)familiarity is here defined as 'socially located feelings of being or not being at home, feelings which are, per definition, not rational, let alone clearly articulated' (Andersen, 2013, p.43). Unlike the authors mentioned above, Andersen is not so much interested in finding out why and when people cross national borders, but in understanding what these borders, and the feelings of (un)familiarity associated with them, mean to the people who either cross them or do not. By claiming that '(un)familiarity ... reflects the individual's sense of self' (p.48) – and by thus translating it into 'self-other relationship' – she apprehends the border experience as an element of narrative identity formation.

As mentioned above, cross-border residential mobility represents a particularly radical form of border crossing. It can be expected that it involves intensive and unforeseen feelings of (un)familiarity, which ask for equally intensive and ongoing processes of storytelling, narrative re-orientation or self-construction in view of what is expected and encountered on 'the other side'. While the act of buying petrol or liquor in the neighbouring country does not necessarily provoke a self-related story, residential mobility certainly does. Moving one's home – beyond a national border at that – can hardly remain untold.[9]

However, our study is of interest not only because it looks into the particularly intensive experiences or narrations of (un)familiarity that accompany the relation between movement, home and identity, but also because it deals with the highly complex and rapidly changing border realities of the Greater Region

SaarLorLux. Although Spierings, van der Velde and Szytniewski emphasize the processual nature of borders and the spatio-temporal complexity of identification processes,[10] the empirical work that informs their conceptual endeavour – especially quantitative data on cross-border shopping (Spierings and van der Velde, 2013) – departs from the notion of individual acts of national border crossing seen as isolated experiences of a comprehensive dichotomy. Our research field, Luxembourg and its various borderlands, is well suited to questioning the idea of a compact dichotomy here/home country/sameness/identity versus there/foreign country/difference/the other that implicitly underlies much of the work on (un)familiarity.

The (un)familiar others in the Greater Region

Our research on residential mobility is situated in the so-called Greater Region or Greater Region SaarLorLux, a loosely defined politico-territorial entity and a space of current economic and socio-political cooperation that is also marked by certain historico-cultural bonds.[11] This entity comprises the German Saarland, the Lorraine in France, the Grand Duchy of Luxembourg – these three being the name givers – and parts of Belgian Wallonia and the German federal state of Rhineland-Palatinate. In our study on residential movement we will not, however, consider the entire Greater Region, but limit ourselves to movements from Luxembourg into the surrounding borderlands and, even more specifically, from Luxembourg to the German side of the common border. Our empirical research, which consisted of open interviews and participant observation in four selected German border villages, was chiefly oriented toward questions of identity construction, i.e. toward experiences and changing notions of cultural similarity and difference, closeness and strangeness (see Boesen and Schnuer, 2015; Boesen, Schnuer and Wille, 2015).

The first particularity of our case is related to geographical size. Luxembourg has borders with Belgium, France and Germany that are equally open, and the small size of the country entails the presence of more than one border at relatively close distance. This means that residential migrants not only decide to settle across the border, but also make a choice between three alternative destination countries. In a more general sense, it implies that experiences of unfamiliarity do not rely only on the encounter with a distinct other, the comparison between us and them, but also on the comparison between various others, between Belgium, France and Germany.

As already mentioned, our study concentrates on German villages in the borderland to Luxembourg that have recently become destinations for residential migrants from Luxembourg. These localities witnessed a considerable population growth due to the influx from across the border, and, not surprisingly, they have likewise seen a significant increase in real estate prices and to a minor degree also in rental prices. However, when compared to Luxembourg, real estate prices are still substantially lower across all three surrounding border regions. This general difference helps us to understand the motivation for departing

Luxembourg but does not explain the particular choice of one of these border regions as the new country of residence over the other two. The decision for a distinct country is based on simultaneous and interwoven experiences of several variants of (un)familiarity. This becomes obvious, for example, when Luxembourgers who have recently settled in a German village are more eloquent in describing their attitudes towards France and the French than those towards Germany and the Germans – a phenomenon to which we will return later.

It is important to note that, in the case under consideration, the encounters with these multiple others and with their more or less familiar characteristics do not necessarily happen in diverse spaces and thus in conjunction with spatial border crossings. With that, we hint at another peculiarity of our research field, namely, the multifariousness and intensity of border crossing activities. The perceptions and experiences of residential migrants are not only linked to their moving to another country but also to various other mobility phenomena, e.g. border-crossing shopping. The inhabitants of the German border region – and equally those of the Belgian and French border regions – buy their petrol, alcohol and tobacco in the Grand Duchy, while shopping centres and supermarkets in Germany attract increasing numbers of customers from Luxembourg. Even more remarkable than these examples of intensive mobile consumerism is the degree of border-crossing labour commuting – although this movement goes almost exclusively in one direction.[12] More than 44 per cent of employees in Luxembourg are cross-border commuters, 25 per cent of which are commuters from Germany who live in the border region.

This high degree of labour commuting – the Greater Region shows by far the highest commuting rates in the EU – helps us understand that the notion of the other as being someone on the other side of the border, that is, the idea of experiences of compact dichotomies – we here and they over there – is misleading. The knowledge Luxembourgish residential migrants possess about 'the Germans' has, in large part, not been acquired in Germany but in Luxembourg, for example at their own workplace, i.e. in encounters with German colleagues, who, after the relocation of residence, might have turned into their neighbours. The other, with whom they are confronted at their new place of residence, is thus in multiple ways part of their hitherto familiar social environment, their lived-in world.

Ultimately, the border crossing activity which most strikingly shows that compact dichotomies have become obsolete is residential mobility itself. While for some mobile individuals, and perhaps also for certain local communities, national territorial borders seem to be dissolving, other examples show that the border can, so to speak, be moved and thus maintained. In several German border villages we now find development areas that are predominately inhabited by Luxembourg nationals and can be described as enclaves, the villages themselves being, as it were, divided by national territorial boundaries. Those who move from Luxembourg into such residential areas are not so much confronted with 'the other' but with '*dem Eigenen*', i.e. with what is proper to themselves (Boesen et al., 2015).

The third particularity of our case of cross-border residential migration is the national diversity of those who relocate their home. This diversity is related to the composition of Luxembourg's population that is comprised of 45 per cent non-Luxembourgers. In recent years, the neighbouring borderlands have become attractive not only for native Luxembourgers but also for members of various migrant groups. For instance, statistical data from 2010 show that 32 per cent of those who have recently moved from Luxembourg to French border towns belong to the group of Portuguese migrants,[13] the demographically largest category of foreigners in the Grand Duchy.[14] The development of the German villages we studied is also revealing in this respect. Over the space of a few years they have turned from distinctly rural localities with an almost exclusively German – and autochthonous – population into, as it were, cosmopolitan communities. For those who move from Luxembourg to the German borderland, this means that they can choose between the aforementioned Luxembourgish enclaves and neighbourhoods with more than thirty nationalities – places where the everyday experiences of its residents are barely influenced by the national divide of Luxembourg-Germany.

Dwelling in (un)familiarity

Given the lack of conceptual clarity regarding the notion of (un)familiarity, we aim to empirically engage with this feature of life in borderlands by looking for accounts of change, differences, comparisons and knowledge that relate both to the place of origin and the residential migrants' new home. We first consider Luxembourgish nationals and then members of the international community that have relocated from Luxembourg to German border towns. It is important to note that the distinction 'Luxembourgish' and 'international' is a simplification and in many ways fragmentary. Both the Luxembourgish nationals and those described as 'internationals' are very heterogeneous and in some instances overlapping groups.

Luxembourgish residential migrants

Cross-border residential migration can, conceptually at least, be split into two processes: 'moving' and 'dwelling'. These are not entirely separate in reality, but the distinction is useful in order to discuss different aspects of the overall phenomenon. The first process – the relocation of residence – is often addressed by the interviewees by narrating how they have come to leave their former place of residence and to live in their current home and their particular village. The second process – dwelling – appears in the data as details about the new place of residence, about neighbourhoods, and about the process of home-making, often featuring particular illustrative anecdotes.

In order to present the experiences of Luxembourg nationals concisely, we are going to focus on one couple in particular, René and Monique Bernard,[15] since they are good exemplars in the sense that they treat many of the issues that others also, and repeatedly, raise. The Bernards, both around sixty years old, have built a small bungalow in Freudenburg, a village some 18 kilometres beyond the border with 1,700 inhabitants. The place has a relatively well-developed infrastructure, featuring a kindergarten and a primary school as well as basic shops, some restaurants and some other small commercial businesses. Perhaps more important in the case of the Bernards is the fact that the village is home to a number of societies and clubs which are active in organizing various festivities and local events.

The Bernards, like many other residential migrants, mention real estate prices, which have risen to unprecedented levels in Luxembourg in the past few years, as one of the main reasons for moving to Germany, where buildable land especially is both available and comparatively affordable. For the Bernards the price increase in Luxembourg was of particular importance, since they had planned to purchase a property for their retirement, but when the time came they found that their savings had fallen behind the inflating prices. 'When the time came [to look for something to buy], when we went to look for apartments, then you threw your hands up in despair with the money that you had saved,' Monique notes. Not only had this aspect of Luxembourg changed, but it has also changed without them being fully aware of it. The interviewees' accounts often contain this kind of historical comparison, with prices not being perceived as merely high, but as having changed dramatically.

The real estate market is not, however, the only facet of Luxembourg as a place that has undergone change. Another often mentioned alteration is the shift towards French as the lingua franca in shops, hospitals and other services. Accounts of this change do not focus on possible causes – the increase in the number of French cross-border workers, for example – but on the experience of alienation, on the irritation of not being able to use one's mother-tongue in one's own country, and also on a sense of frustration with the perceived rudeness that has to be endured in these instances of alienation – the rude insistence on French by waiting staff in cafés and restaurants is an almost universal feature in the interviews.

France and the French also appear in other contexts. As previously mentioned, the decision to move across the border is not a dichotomous choice between places A and B, since there are several borderlands to be chosen from, all of them marked by a difference in real estate prices but each with its own particular characteristics. Asked whether they had also considered moving to either France or Belgium, many Luxembourgers in the German borderland answered in the negative and were particularly explicit about the fact that they could not imagine living in France. The decision to move to Germany is often a mixture of a positive and a negative choice – opting for Germany also comprises opting against France. National comparisons are thus made not only between Luxembourg and Germany, but also between France and Germany. The latter country is distinguished both as tidy and well-organized, yet also as culturally and especially linguistically more similar. 'The French', both those across the border from Luxembourg in France and those within Luxembourg, are, in many ways, regarded as 'the other'. The narratives of residential mobility do not only reveal a complex set of differences and parallels across national borders, but are interwoven with the process of living in and through a changing place, a changing home, that is becoming less familiar.

The residential migrants are not only moving away from Luxembourg, they are also dwelling in their new home. The Bernards have built their new home in an area that they were previously only familiar with in the most superficial sense – that is, knowing it exists and having visited parts of it infrequently. Unsurprisingly, there are many 'unknowns' that must be recognized and 'lived-with' as part of the process of dwelling in the new place. One such unexpected feature is the degree of similarity between the local German dialect and the Luxembourgish language. The Bernards, like other Luxembourgish residential migrants, mention how they spoke standard German at the local shop and with their neighbours in an effort to integrate and be polite, thereby implicitly denouncing the indifference in this regard that they experience in Luxembourg. It was only upon the insistence of the autochthons to speak Luxembourgish that the Bernards began to use their mother tongue in their new place of residence. 'For example, at the bakers here, I spoke German out of politeness, and then she said "You can simply speak Luxembourgish with me." "Oh," I replied, "how lovely, I can speak Luxembourgish here, but at home in my own country I can't."' For the Bernards this offer was the expression of an unexpected familiarity sharply contrasting with the alienation experienced at home. Life in the German village is different from life in today's Luxembourg by being similar to the way that life in Luxembourg used to be.

Equally notable is how the Bernards, like some of the interviewees, describe the ease with which they forge social relationships. The only thing required in this natural process is openness and the willingness to participate. Many of the interviewed Luxembourgers describe the 'mentality' of the locals with regard to social interactions and neighbourliness as unexpected and unfamiliar. This neighbourliness is the subject of frequent comparison with Luxembourg in the interviews; the new neighbours are said to be 'more social' than their former neighbours in Luxembourg. It is this particular aspect of the place that appears different and unfamiliar, yet at the same time permits and facilitates the process of familiarisation. Often this familiarity established through social relationships with neighbours and participation in town life is central to the personal contentment of the newcomers.

Whereas Luxembourg is frequently described as having changed in a way that makes it increasingly unfamiliar and gives people like the Bernards the feeling of being 'out of place', the narratives of dwelling in the new place mirror converse experiences: the migrants discover familiarities and they get to know differences that become familiar and valued parts of dwelling at their new home. Familiarity is thus not a state of being, but part of the process of dwelling.

International residential migrants

The heterogeneity of the group of residential migrants was mentioned above. In addition to Luxembourgish citizens, it comprises a large sub-group of internationals, i.e. people who were already migrants in Luxembourg, often working for international banks, insurance companies or European institutions. Obviously the relationship of these people to Luxembourg as a dwelling place is different from that of the nationals. It can be presumed that this difference will also recur in the accounts of relocating across the border. In order to discuss the group of international migrants, it is preferable not to present one particular family; instead we will present the situation in Auf Mont, one of the residential developments in the region situated in the Moselle village Wincheringen. The development is distinct in several aspects.

First, Auf Mont is an exceptionally large development, with more than three hundred residential lots being planned. Second, in contrast to other cases where the local municipality purchases the land, parcels it up, constructs roads, etc. and eventually sells the fully developed plots, this residential project is completely in the hands of a private investor. This entrepreneur had a coherent and detailed vision of the neighbourhood to be created, which features clearly in the advertisement message of the 'Family Park Auf Mont,' as it is called: 'One thing that makes AUF MONT so special is the variety of nationalities and various stages of life of our neighbors. Young couples, planning to start a family are living next to families that desire to spend their sunset years in a wonderful neighborhood' (Auf Mont Family Park). This international and heterogeneous community is quite clearly sought in Luxembourg, where advertisements for the Family Park frequently appear in newspapers and magazines. The official website emphasizes the short commute to Luxembourg and underlines the scenic setting and the presence of facilities such as the multilingual kindergarten, but it provides no information about (and barely a mention of) the village of Wincheringen, at the edge of which it is located. The advertisements evocatively present a neighbourhood that somehow resembles a Club Mediterranée colony, and they give the impression that, except for the short distance to Luxembourg, this development could be located anywhere.

In many ways, Auf Mont has been a success. It is home to international couples and young families and so is steadily developing into the very community it aspires to be. The narratives of moving delivered by our interviewees who live here are different from those of Luxembourgish families like the Bernards. They too refer to real estate prices in Luxembourg as a key motive, but the prices are described as high, not as having increased intolerably. Their decision to move was the final outcome of a long search in Luxembourg and the realisation that building or buying a house there was economically irrational. The fact that more emotive accounts revealing a progressive estrangement from Luxembourg are absent shows that the initial relationship of this group to Luxembourg as a place of dwelling is different. The internationals living in Auf Mont have a much more 'synchronic' relation to their former home, focusing on the present and largely ignoring the ways in which Luxembourg has changed. Without a deeper attachment to Luxembourg and lacking the experience of its former condition, they feel neither the regret nor the relief that are expressed by native Luxembourgers now settled in Germany. Living amid people who are in many respects like themselves, the international residents in Auf Mont have no acute feeling of having left a country. The colony appears to them as a kind of extraterritorial bulge or as a suburb of the metropolitan, international Luxembourg with which they are acquainted and which alone has relevance for them. Auf Mont is a place where they find each other to be neighbours, looking for communal contact which, unlike in the case of the Bernards, is forged apart from the village life. Having an international background is considered a similarity and a uniting factor. On the other hand, since Luxembourgers have also moved to the settlement, it is here, paradoxically, that many internationals get acquainted with an appreciable number of Luxembourg nationals, while the previous neighbourhood in Luxembourg, when compared with the newly acquired one, is generally said to have been anonymous and reserved.

Auf Mont provides us with a different version of how a national dichotomy cannot adequately capture (un)familiarity. Whereas the Luxembourgish family illustrated how the dichotomy is broken up by the changing Luxembourg, affected by cross-border labour mobility amongst other things, the international community in Auf Mont shows how familiarity need not necessarily denote homogeneity, that is, sameness, but can also describe a heterogeneous group that seems to exist beyond the national dichotomy that plays out around them. These multicultural cosmopolitans were attracted by something ostensibly familiar and recognisable, an image of a suburban idyll that afforded them the possibility of building their own home and living with others who they consider to be interesting and different, yet like-minded.

Conclusion

Our discussion of the literature focused on the fact that in the work of van der Velde, Spierings and Szytniewski the concept of (un)familiarity is often used interchangeably with difference and similarity. In the same vein, in a given border situation, the 'home side' is regarded as familiar and 'same', and the 'other side' as (un)familiar, in the sense that it is somewhere between different and similar to the 'home'. Our empirical data show these relations to be more complicated than that. In studying Luxembourgish residential migrants, we encountered complex relationships with more than one 'other side' being important for moving and dwelling. We also saw that difference can be a source of familiarity, in the sense of 'feeling at home', while (un)familiarity, in the sense of a mixed feeling of familiarity and unfamiliarity, does not necessarily result from difference. We also noted that, importantly, becoming familiar with a place, such as a new neighbourhood, and being familiar with a place, are each processes that are ongoing. Familiarity is not an achieved status, but it is the maintenance of feeling familiar. Rather than 'home' being given as familiar, dwelling in this 'home' is familiar as a consequence of continuous encounters with the place and (some of) its inhabitants.

Dwelling and the familiarity of the residential migrants with their new home thus involves participating in a place on an everyday basis – greeting neighbours, dropping children off at school, maintaining relationships. It comprises getting to know other residents, perhaps joining clubs and societies, being aware of the social calendar, etc. Familiarity, in this case, does not necessarily rely on comparative similarity; differences, too, become a source of feeling 'at home'. This can be clearly seen by looking back at the ingredients of the familiarity the Bernards feel: on the one hand, a language that is unexpectedly similar to their own; on the other hand, neighbourliness that is surprisingly different from expectations and previous experiences. That difference and familiarity can go together is also apparent when looking at cosmopolitan residential migrants. In Auf Mont, where most people are migrants from a wide variety of countries, the national difference/sameness lost almost all of its relevance. Instead it is the 'sameness through difference' of its residents that takes precedence, and it is exactly this shared difference that makes living in Auf Mont familiar and attractive.

In these instances of dwelling, where differences can be seen to be a source of familiarity, we can also reconsider proximity. We have shown that proximity, in the more abstract understanding of a feeling of nearness, is difficult to distinguish clearly from similarity and, by the same token, from familiarity – what is a feeling of nearness and closeness other than a feeling of familiarity? What about proximity in the more literal sense of living together in a place, that is, participating in its everyday life, whether this involves direct interaction or not? We would argue that this literal conception of proximity can be a more promising approach for understanding the familiarity of dwelling in a place – an understanding that does not rely on the notion of similarity/difference. In Auf Mont the continued presence of diverse nationalities is part of the daily physical and interactional proximity with and within a place. Similarly, the participation in distinctly local and hitherto unknown neighbourly encounters, described by the Bernards and other Luxemburgish migrants, as well as the continual interchange with the autochthones in their, as it were, common mother tongue are equally part of this lived, concrete proximity.

This chapter has argued against the (implicit) equation of familiarity/unfamiliarity with sameness/difference and has tried to show that the confusion of both might be caused, at least in part, by focusing on mobility and neglecting the process of dwelling – or, to put it differently, by the neglect of the relevance of proximity, of being in the same place, to feelings of (un)familiarity.

Notes

1 See particularly 'Regulation (EU) No 492/2011' on the freedom of movement of workers and 'Directive 2004/38/EC' on the right of citizens of the EU and their family members to reside and move freely.
2 See Jagodic, 2012 for an introduction to other examples; see also Strüver, 2005a, 2005b; van Houtum and van Naerssen, 2002; Gielis and van Houtum, 2012.
3 See, for example, van der Velde, 2012; van Houtum and van der Velde, 2004.
4 De Fátima Amante, 2013 and Andersen, 2013 as well as the contributions to a *European Planning Studies* special issue (January 2013) and a *Journal of Borderland Studies* special issue (3, 2014); see also the European Science Foundation financed research project on 'Unfamiliarity as signs of European times: scrutinizing historical representations of otherness and contemporary daily practices in border regions'; available online at www.unfamiliarity.eu/.

5 See, i.a., Spierings and van der Velde 2008, p.503: 'Borders are necessary to sustain a certain level of "unfamiliarity", a "desired difference" so to say.'
6 Although the term (un)familiarity had not been introduced at that stage, see also van Houtum and van der Velde, 2004.
7 See Schiffauer, 2004; 2014.
8 Mayol's description of the urban *quartier* and its 'art de coexister avec des partenaires (voisins, commerçants) qui vous sont liés par le fait concret, mais essentiel, de la proximité et de la répétition' (de Certeau et al., 1994, p.17) comes to mind. See, however, recent attempts at overcoming this narrow concept of dwelling: e.g. Stock on 'polytopisches Wohnen' (Stock, 2009).
9 See in this context Metcalfe, 2006; Leach, 2002; Rapport and Dawson, 1998.
10 See particularly Szytniewski and Spierings, 2014.
11 For further information on the nature and constitution of the Greater Region, see Wille, 2008.
12 In 2011 more than 145 000 people commuted on a daily basis to Luxembourg, whereas less than a thousand people left the country for their workplace in one of the three neighbouring countries; Interregionale Arbeitsmarktbeobachtungsstelle, 2012.
13 See Carpentier, 2010.
14 More than 16 per cent of the Luxembourgish population are of Portuguese nationality; Statec, 2014.
15 All names are anonymized.

Bibliography

Andersen, D. J., 2013. Exploring the Concept of (Un)familiarity: (Un)familiarity in Border Practices and Identity-Formation at the Slovenian-Croatian Border on Istria. *European Planning Studies*, 21(1), pp.42–57.

Auf Mont Family Park. Available online at: www.aufmont.com/en/8/what-is-auf-mont/ [Accessed 15 October 2014].

Bhatti, A. and Kimmich, D., 2015. Einleitung. In A. Bhatti, and D. Kimmich, eds., 2015. *Ähnlichkeit. Ein kulturtheoretisches Paradigma*. Konstanz: Konstanz University Press, pp.7–30.

Boesen, E., Nienaber, B., Roos, U., Schnuer, G. and Wille, C., 2015. Phantomgrenzen im Kontext grenzüberschreitender Wohnmigration. Das Beispiel des deutsch-luxemburgischen Grenzraums. *Europa Regional*, 22(3/4), pp.114–128.

Boesen, E. and Schnuer, G., 2015. Wohnen jenseits der Grenze. Regionale Integration und ihre lokale Verwirklichung. In C. Wille, ed., *Lebenswirklichkeiten und politische Konstruktionen in Grenzregionen. Das Beispiel der Großregion SaarLorLux. Wirtschaft – Politik – Alltag – Kultur*. Bielefeld: Transcript, pp.179–201.

Boesen, E., Schnuer, G. and Wille, C., 2015. Urbanität im ländlichen Raum. Wohnmigration in der deutsch-luxemburgischen Grenzregion. In R. Garstenauer and A. Unterwurzacher, eds., 2015. *Aufbrechen, Arbeiten, Ankommen. Mobilität und Migration im ländlichen Raum. Jahrbuch für Geschichte des ländlichen Raumes 2014*. Innsbruck: Studien Verlag, pp.225–244.

Carpentier, S. ed., 2010. *La mobilité résidentielle transfrontalière entre le Luxembourg et ses régions voisines*. Luxembourg: Editions St Paul.

Certeau, M. de, Giard, L. and Mayol, P., 1994. *L'invention du quotidien, t.2 habiter, cuisinier*. Paris: Gallimard.

De Fátima Amante, M., 2013. Recovering the Paradox of the Border: Identity and (Un) familiarity Across the Portuguese-Spanish Border. *European Planning Studies*, 21(1), pp.24–41.

Faist, T., 2000. Transnationalization in international migration: implications for the study of citizenship and culture. *Ethnic and Racial Studies*, 23(2), pp.37–41.

Gielis, R. and van Houtum, H., 2012. Sloterdijk in the House! Dwelling in the Borderscape of Germany and the Netherlands. *Geopolitics*, 17(4), pp.797–817.

Interregionale Arbeitsmarktbeobachtungsstelle, 2012. Die Arbeitsmarktsituation in der Großregion 8. Bericht der Interregionalen Arbeitsmarktbeobachtungsstelle an den 13. Gipfel der Exekutive der Großregion, Saarbrücken.

Jagodic, D., 2012. Living (beyond) the Border: European Integration Processes and Cross-border Residential Mobility in the Italian-Slovenian Border Area. In D. Jagetic Andersen, M. Klatt and M. Sandberg, eds., 2012. *The Border Multiple: The Practicing of Borders between Public Policy and Everyday Life in a Re-Scaling Europe*. Furnham: Ashgate, pp.201–218.

Klatt, M., 2014. (Un)familiarity? Labor related Cross-Border Mobility in Sonderjylland/ Schleswig since Denmark joined the EC in 1973. *Journal of Borderland Studies*, 29(3), pp.353–373

Löfgren, O., 2008. REGIONAUTS: The Transformation of Cross-Border Regions in Scandinavia. *European Urban and Regional Studies*, 15(3), pp.195–209.

Leach, N., 2002. Belonging: Towards a Theory of Identification with Space. In J. Hillier and E. Rooksby, eds., 2002. *Habitus: A Sense of Place*. Aldershot: Ashgate, pp.281–295.

Metcalfe, A., 2006. "It was the Right Time to do It": Moving House, The Life-course and 'Kairos'. *Mobilities*, 1(2), pp.243–260.

Rapport, N. and Dawson, A., 1998. Home and Movement: A Polemic. In N. Rapport and A. Dawson, eds., 1998. *Migrants of Identity. Perceptions of Home in a World of Movement*. Oxford: Berg, pp.19–38.

Schiffauer, W., 2004. Cosmopolitans are Cosmopolitans: On the Relevance of Local Identification in Globalizing Society. In J. Friedman and S. Randeria, eds., 2004. *Worlds on the Move*. London; New York: I.B. Tauris, pp.91–101.

Schiffauer, W., 2014. Migration und deutsche Identität. Deutschland braucht "inklusives Wir-Gefühl". *Deutschlandradio Kultur*, 21.1.2015.

Spierings, B. and van der Velde, M., 2008. Shopping, Borders and Unfamiliarity: Consumer Mobility in Europe. *Tijdschrift voor Economische en Sociale Geografie*, 99(4), pp.497–505.

Spierings, B. and van der Velde, M., 2013. Cross-Border Differences and Unfamiliarity: Shopping Mobility in the Dutch-German Rhine-Waal Euroregion. *European Planning Studies*, 21(1), pp.5–23.

Statec, 2014. *Luxemburg in Zahlen*. Luxembourg: Institut National de la Statistique et des Etudes Economiques.

Stock, M., 2009. Polytopisches Wohnen – ein phänomenologisch-prozessorientierter Zugang. *Informationen zur Raumforschung*, 2009 (1/2), pp.107–116.

Strüver, A., 2005a. Spheres of Transnationalism within the European Union: On Open Doors, Thresholds and Drawbridges along the Dutch-German Border. *Journal of Ethnic and Migration Studies*, 31(2), pp.323–343.

Strüver, A., 2005b. *Stories of the 'Boring Border'*. Münster: LIT.

Szytniewski, B. and Spierings, B., 2014. Encounters with Otherness: Implications of (Un) familiarity for Daily Life in Borderlands. *Journal of Borderlands Studies*, 29(3), pp.1–13.

Van der Velde, M., 2012. Boring European Borders?! Integration and Mobility across Borders. *Eurasia Border Review*, 3, pp.115–126

Van Houtum, H. and van Naerssen, T., 2002. Bordering, Ordering and Othering. *Tijdschrift voor Economische en Sociale Geografie*, 93(2), pp.125–136

Van Houtum, H. and van der Velde, M., 2004. The power of cross border labour market immobility. *Tijdschrift voor economische en sociale Geografie*, 95(1), pp.100–107

Wille, C., 2008. Grenzgänger zwischen Regionalität und Globalität. Spuren kultureller Flexionen in der Großregion SaarLorLux. In W. Amann, G. Mein and R. Parr, eds., 2008. *Periphere Zentren oder zentrale Peripherien? Kulturen und Regionen Europas zwischen Globalisierung und Regionalität*. Heidelberg: Synchron-Verlag, pp.47–60.

Yndigegn, C., 2013. Reviving Unfamiliarity – The Case of Public Resistance to the Establishment of the Danish-German Euroregion. *European Planning Studies*, 21(1), pp.58–74.

10 The residential and symbolic dimensions of cross-border mobility

Looking at members of the French middle class in the agglomeration of Lille

Garance Clément

Introduction

European borderlands offer a special window on different research topics such as the European enlargement process, local governance and identity construction (Bauer and Darley, 2007). Accepting the axiom that 'some things can only occur at borders' (Donnan and Wilson, 1999, p.4, cited in Jagodic, 2014), the study of borderlands can also enrich the understanding of everyday life practices (Bolzman and Vial, 2007) and residential behaviours (Carpentier, 2012; Boesen, Schnuer and Wille, 2015). In many respects, borderlands appear as lands of opportunities for border populations. While the processes of border opening and policy integration have made cross-border mobility easier, national control of tax systems and urban planning maintain important asymmetries in real estate prices, prices of daily products and taxes. Depending on resources, life trajectories and habitus, people can benefit from this paradoxical evolution of borderlands in various ways. One of these ways that has not received much scholarly interest yet is 'cross border residential mobility' (Carpentier, 2012) or 'cross border residential migration' (hereafter CBRM).

In this chapter I focus on this specific form of spatial mobility which differs from long distance migration in that it creates situations where people are 'sleeping abroad but working at home' (Balogh, 2013). The singularity of this mobility lies in the fact that the search for a new home induces a long-term change of legal status – from national to foreign resident. When they settle on the other side of the border, 'cross border migrants' experience new rules and norms. They develop new lifestyles in which they have to cross the border every day in order to maintain their previous links. Research on CBRM has investigated this tension between 'being abroad' and 'being near home' from the perspective of transmigration and border studies, and sometimes with attempts to link the two approaches (Gielis, 2009; Balogh, 2013). This chapter is not so much about how people experience borders and border definitions than it is about how borderlands are specific contexts that reveal social trends. I will draw on theoretical work from urban sociology (Grafmeyer and Authier, 2008) and the branch of migration studies that looks at 'lifestyle migrations' (Benson and O'Reilly, 2009). I argue that they offer new insights on what is at stake for border populations who engage in CBRM. My goal is to explore the logics behind cross-border residential mobility by focusing on the social determinants and conditions that allow for it to happen, as well as the strategies that it encompasses. I look at borders as social constructs which have legal and material dimensions and may constitute a resource for borderland inhabitants. Moreover, CBRM is brought into the scope of research that views international or cosmopolitan experiences as a form of capital that matters more and more in the contemporary world (Wagner, 2007). In this perspective, looking at borderlands and the residential strategies that they entail provides us with the possibility of investigating both international and local dimensions of social classes.

Empirically, the chapter will use the case of the residential mobility of members of the French middle class who leave the agglomeration of Lille and settle a few kilometres away, across the Belgian border, while maintaining their professional activity in France. In 2006 more than 25,000 French residents were settled in the Belgian borderland, mainly in the Walloon region, where they represented 7 per cent of the population (ADULM, 2010). In Mouscron, one of the largest Belgian border cities, this proportion reaches 22 per cent of the population (SPF Économie, 2011). Franco-Belgian cross-border migrations are not a new phenomenon. On the contrary, this border has a long history of intense contact between border populations. This tradition goes back to industrial developments of the nineteenth century when the need for labour in textile industries in France created a significant influx of workers coming from Belgian border territories.[1] Today's flows are reversed, as many French factory workers seek jobs in the Belgian borderlands where unemployment rates are lower and wages higher. But aside from these long-established, job-driven migrations, another kind of mobility has much increased in the last two decades, namely the cross-border residential mobility of members of the middle and upper classes. Given the absence of data on this phenomenon so far, the chapter will offer an analysis of qualitative data gathered through interviews with middle class migrants[2] and local actors who take part in or witness this new kind of migration. The chapter has five sections. The first two sections of the paper further develop the theoretical background and the empirical context of the study. The following sections explain the social determinants of Franco-Belgian CBRM, provide elements of the strategies that this mobility entails, and offer conclusive insights into its symbolic and material dimensions.

Bringing CBRM into the scope of urban sociology and migration studies

So far, aside from a few works in social sciences (Bolzman and Vial, 2007; Boesen, Schnuer and Wille, 2015; Carpentier, 2012; Rérat, et al., 2011) scholars who have studied CBRM have done it with the ambition of understanding the way mobility transforms the experience and reality of borders, either to argue in favour of transnationalism (Jagodic, 2014) or to demonstrate processes of (re) bordering, ordering and othering (Balogh, 2013). Urban studies associated with some findings from lifestyle migration studies offer a complementary perspective on CBRM. Urban sociology explores the social trends emerging from a specific urban context (Grafmeyer and Authier, 2008). Since the 1960s, urban sociology has contributed to the analysis of urban issues and has notably helped in understanding the impact of globalization on cities (Le Galès, 2005; Sassen, 2011). European border cities are particularly interesting research fields because they combine all the traditional elements of urban life, allowing researchers to investigate suburbanization beyond administrative borders as well as state borders. Within French urban sociology, the scholars who look at residential trajectories and urban lifestyles have not paid much attention to border cities as specific environments. Yet cross border contexts provide good case studies for a number of growing fields of interest in urban sociology, especially regarding the spatial organisation of social classes (Oberti and Préteceille, 2004) and the relationship between residential and social mobility (Maurin, 2004; Dietrich-Ragon, 2013; Cusin, 2012). Indeed, the study of cross-border residential mobility and the way people formulate their housing choices can help in understanding processes of downward and upward social mobility. This form of mobility reveals phenomena of segregation that can relate to voluntary strategies of self-exclusion among higher classes, or to mechanisms of domination over the poorest. When it comes to the middle class, residential mobility is often analysed from the perspective of gentrification to show how the residential ambitions of people with higher cultural or economic capital transform a poorer neighbourhood (Collet, 2012a; Clerval, 2013). The analysis of middle class residential mobility also accounts for phenomena such as urban sprawl and suburban lifestyles (Haumont, 1966; Charmes, 2005). In this chapter, the focus is on the way residential mobility, when it takes place in an international context, redefines social position.

This kind of analysis is also developed in research on lifestyle migrations. Lifestyle migration studies put the emphasis on 'the relocation of people within the developed world searching for a better way of life' (Benson and O'Reilly, 2009, p.608). It is an international migration that modifies migrants' legal and symbolic positions. Benson and O'Reilly contend that lifestyle migration should not only look at traditional categories of expatriates but also at 'relatively affluent migrants who do not fit the stereotypical idea of a migrant in the given destinations' (Benson and O'Reilly, 2009, p.609). The French middle class people moving to Belgian border cities, who are my subjects, engage in migration over a very short distance, but they nevertheless seem to belong to this growing trend of relatively affluent people looking for more fulfilling places to live. Lifestyle migration studies are not only concerned with migrants' projects for better living but also with a 'narrative of escape'. This narrative can be found to some extent in the discourse of the French middle class representatives in question. Another interesting aspect about lifestyle migration studies is that, relying on Bourdieu's work, they pay attention to the way migrants' lifestyles can be traced back to class habitus. Like Anne-Catherine Wagner in her work on social classes in France, lifestyle migration scholars study migration and mobility not as isolated events, but as part of the lifestyle trajectories of migrants. They relate lifestyle to class, using the Bourdieusian concepts of habitus, dispositions and tastes (Oliver and O'Reilly, 2010). This approach helps understanding how migration can be part of a process of social distinction (Benson, 2013).

Both urban sociology and lifestyle migration studies contend that local practices derive from and (re)produce social distinctions. They share a special interest in the spatial dimension of social domination mechanisms by looking at forms of segregation that are produced in urban areas. Finally, they pay attention to the processes of socialization that can account for the habitus and the social position of people, including socialization through residential and urban space. This combination of approaches can help clarify and interpret the multiplicity of situations that characterize affluent long-distance migrants (Gellereau, 2014), but it also improves our understanding of local and cross-border migrations. This theoretical background leads me to formulate several questions regarding CBRM. How does the presence of a border in an urban context impact residential careers and social paths? What social and residential strategies drive people who leave one country for another? How does a borderland constitute a resource for inhabitants? Before analysing the case of middle class French residential migrants, it is necessary to better understand the Franco-Belgian context.

The Lille agglomeration and its Belgian surroundings

The Franco-Belgian border is an interesting case to study CBRM because it embodies several conditions that ease borderlanders' mobility. First, this national border is a recent one (it took its final form in 1839 after Belgium's independence from the Netherlands) and its history is marked by intense contacts between border populations, seen, for example, in the large number of mixed marriages (Hamez, 2004). This history still nourishes collective representations of cultural closeness and a community of origins. This notion of closeness is reinforced by discourses produced by local and European institutions involved in cross-border projects, who insist on the proximity of Lille and Belgian urban areas. The fact that these urban areas are contiguous, that there is barely any physical barrier to manifest the border and that France and Wallonia share a common language also explains why there are strong bonds between Belgian and French borderlanders.

The Lille agglomeration is a heavily populated area (1.1 million inhabitants), characterized by strong social and economic inequalities that are observable in various forms of social and spatial segregation (Miot, 2012). The city of Roubaix, which is close to the border, stands out for being the poorest city of France, with an unemployment rate of more than 30 per cent in 2012 for people aged from 15 to 64.[3] At the same time, cities belonging to the 'BMW triangle' (Bondues-Marcq-

Wasquehal) have for decades been among the richest, attracting skilled workers and real estate investors (Paris, 2002). Alongside this segregation between the wealthiest and the poorest, there are also a wide range of people from the middle class for whom housing can have decisive effects on their social position. They mostly occupy the rented sector and lack the resources to become homeowners. Indeed, since 2000, prices and rents have increased faster than incomes (ADULM, 2012). Only the wealthiest can become new property owners in city centres, sometimes joining poor families who already occupy ancient and sometimes squalid homes. The middle class residents have to move away from centres to more rural areas in order to buy a house. Although doing so may fit their desire for a quieter and greener life, it often means that they have to accept longer commutes from their homes to places where jobs are available. At the same time, across the border the housing market is less tight, for economic and demographic reasons but also because of the greater availability of land resulting from different planning regulations. As a consequence, while French cities are experiencing a significant housing crisis and lack affordable dwellings, Belgian border cities offer cheap city homes. The kind of properties that are available also fit French households' desires for a garden and more living space. Most of those who respond to the supply of housing are families with young children.

Another thing that needs to be mentioned in order to understand the context of this case of CBRM is that the transportation infrastructure around the border is very favourable to people living in Belgian border cities and working in the Lille agglomeration. Although there are few public transportation services that cross the border, high-speed roads are easily accessible and remain free in Belgium for all drivers. It is thus quicker to cross the border in Mouscron to work in Lille than to come to the city from rural towns on the French side, where traffic jams coming into Lille are a daily occurrence.

Finally, whereas previous legislation on cross-border workers established taxation rules that favoured middle class households who worked in Belgium and lived in France, the adoption of a binational convention on this matter and the suppression of the status of frontier workers made commuting more advantageous to those who live in Belgium and work in France. Before this 2008 reform, except for few individuals with large fortunes who have historically escaped French taxation in Belgium, civil servants were the only households with true fiscal incentives to settle to Belgium, which explains why they are overrepresented in my material (see next section). From 2008 on, other households could benefit from new taxation conditions, which can partly explain why Belgian municipalities witnessed an increase in French residential registrations – even though at that time the financial crisis lowered real estate prices on the French side of the border as well.

Early border socialization

My fieldwork shows that households usually present CBRM as a smart choice between opportunities made with calculation and intuition. However, an analysis of their discourses helps to identify features of early border socialization that seem to play a role in the decision. Socialization is the process whereby individuals acquire norms and values, behaviours and skills in accordance with their social position. I use the term 'border socialization' to refer to all the experiences that make the border a familiar space for borderlanders. These include adolescent experiences of crossing the border, which is a very common practice among French youth and seems to be particularly important and influential among those who have migrated. Festive events in Belgium and regular nights out in Belgian bars or clubs have structural effects on representations of the border and the neighbour country as being both a familiar and adventurous territory. This is illustrated by the following excerpt from the interview transcript of a 56-year-old woman with parents of Maghreb origin. She lives in Mouscron, where she bought a house with her husband and their child. She is unemployed, and her husband works at a post office.

> **Mrs J.:** I was born in Roubaix, but of course the border always attracted us, the teenagers … Because there were many clubs, a lot of cultural stuff […] We never felt the border in fact. Even though at the time there were customs and so on. There was an easy way in and out. During our youth, we flirted with borders, so we have knowledge of it. For sure it made us want to move towards Belgium. Even to settle there. Even though back then I had no clue I would live in Belgium one day.[4]

This border socialization also includes family traditions of living near the border, especially when the family is binational, as a result of mixed marriages. Cross-border family links make it possible to acquire a high degree of knowledge on socio-cultural and legal norms from the other side of the border. This transmission of knowledge through marriage and family ties makes migration easier. Like adolescent mobility and festive habits, it is part of a larger process of socialization that makes people more open to the idea of expatriating to Belgium. During their professional lives, the middle class migrants I interviewed had 'updated' this knowledge about 'the other side' with more practical and political information, as well as relating representations of the border. Public workers in particular have internalized the discourse of local institutions on a 'territory without borders' or a 'cross border region'. Professional life is also the period during which people become more acquainted with fiscal matters and housing market realities. In public administrations on the scale of a city, agglomeration or region, knowledge is shared on a variety of issues such as housing markets and the conditions that need to be fulfilled to become a Belgian resident. Several interviews also revealed that people who have settled in Belgian border cities can be identified in car parks by their licence plates and are sometimes approached in order to get details about their expatriation. These cross-border experiences and modes of knowledge acquisition can be decisive in residential trajectories. In the Bourdieusian perspective, socialization – or sociogenesis – is responsible for the development of habits and representations that constitute the habitus (Darmon, 2010). Indeed, the Bourdieusian definition of habitus refers to a system of structured and structuring dispositions incorporated by individuals during education. Border socialization is one of the elements that take part in the constitution of the habitus of the middle class that I focus on.

Middle class habitus and residential choices

Interviewed households belong to the middle class according to criteria of professional occupation, income and social background. It is possible to distinguish two groups among these households. The first and largest one comprises mainly employees of the public sector or teachers, with a smaller share of junior managers in the public and private sector. They fit the description of middle class given by Catherine Bidou-Zachariasen (1984), inspired by American authors like A. W. Gouldner, in which incumbents of emerging professional positions in the public, cultural and educational sectors in particular opt for larger and cheaper accommodations in suburbs where they can take part in an active local life. In this first group, people have modest backgrounds but not poor ones, which can be seen in the following circumstances: none of them had lived in social housing; they all went to school and have a degree, even if it is a low one; their parents managed to become homeowners; and very few lived in apartments.

The second group is made up of migrants who occupy lower positions in the public sector or are small storekeepers and craftsmen. They have fewer economic and cultural resources than the members of the first group and belong to what some authors have called the ‘modest means’ (Cartier et al., 2008). They come from poorer families, with parents often originating from the Maghreb or Southern Europe, who arrived in Lille as industrial workers during the 1950s.

All migrants from both these groups have achieved a small amount of social upward mobility. They are more educated and have jobs that rank higher in the professional hierarchy than those of their parents. Interviews show that the quite similar social position that migrants occupy is not only observable in terms of earnings and professional occupation but also in the way they make their residential choices. For all migrants, CBRM has been driven by the desire to improve their residential position both in terms of housing conditions and living environment. Migrants can access cheap houses of relative high quality in neighbourhoods with good reputations and numerous facilities that match the tastes and consumption practices of the middle class. These residential choices account for individual habitus but also a middle class habitus which constitutes a system of representations and tastes that is common to the people occupying similar positions in the social hierarchy. In the case of the Franco-Belgian border, the fact that migrants look outside of the country for housing solutions seems to confirm the hypothesis developed by Eric Maurin (2004) that the middle class is characterized by a ‘fear of relegation’, and that housing – and property ownership in particular – is one factor that the middle class uses to secure or improve its social position. I now describe these residential choices for the two groups identified earlier, who belong to the middle class but have different amounts of resources. Migrants with lower economic and cultural resources depict Belgian cities as an Eldorado for housing that offers an opportunity to leave behind the strong constraints that they experienced in France. Their mobility is driven by the desire to become homeowners, something they could not do in France without significant sacrifices.

> When I look at the price of my house compared to the French prices … I bought it for 150,000 euros, I got in, I moved my furniture in and that was it … There was some decoration work to do but without this opportunity, I would never have been able to buy in France. It was impossible, considering the market prices, it is impossible…
>
> (Mr L, civil servant, 35 years old, born in Tourcoing to Algerian parents, left France in 2006 to buy a house in Mouscron with his wife, also a civil servant)

Having witnessed their parents' economic difficulties, these migrants have looked for solutions to escape from them.[5] They are more attracted by city centre accommodation, mainly brick townhouses with a backyard, also known as '1930s houses', and by environments that offer common facilities for daily needs, health and social services and transportation infrastructure, while being easily accessible by car from Lille and its immediate surroundings. The strategy of these migrants is to access property in an area where housing is both cheaper and nicer, avoiding the unpleasant choice between social housing and a significant distance from city centre facilities. It is a way to confirm an intergenerational social mobility that is observable through professional positions. However, these strategies are not always successful and durable.

More affluent migrants, for their part, feel drawn to the quality of the environment that Belgian cities offer. The fact that they are able to buy a house there confirms their social position, matching the position of their parents rather than upgrading it. They choose houses in city centres but also in their close peripheries, where they find more recent constructions in former rural areas where land is cheap. They sometimes build their houses themselves. Interestingly, interviews show that they do not necessarily try to decrease their housing expenses, but decide to buy better accommodation than what would have been affordable in France for the same amount. They often underline the fact that they would have been able to buy a house in France but they also note the restrictions their budget would necessitate, including the different quality of neighbourhood and infrastructure to which they would have access. The fact that they moved out of France to become owners is not presented as a confession of a lower social position but rather as something to be proud of, as I explain in the next section. These migrants justify their choice by praising the proximity of rural areas, the presence of large green spaces and the calm and cleanliness of Belgian cities. The following excerpt underlines the differences between the living conditions in France and Belgium by referring to a more appealing 'Belgian lifestyle'.

> **Me:** When you say you like the 'Belgian lifestyle' … what do you mean exactly?
> **Mrs S.**: Well I think that … for one thing it is clean. I don't know … in Mouscron we are in an area that's … a bit of the countryside while still being in town. So this is also something I like.
>
> (Mrs S., a 40-year-old woman who runs a hair salon and built a house with her husband in Mouscron)

Other aspects worthy of mention are the school system and educational amenities. The middle class is very concerned with the educational trajectories of their children and adopts various strategies when it comes to choosing a school for them. According to Agnes van Zanten (2009), the different social groups that constitute the middle class see school as a way towards autonomy, an investment in the future and a place where children express themselves and achieve self-fulfilment. For migrants, moving to Belgium should not put these achievements at risk. During the interviews, some of them take pride in the fact that their children can enter Belgian schools that have good reviews and join local Belgian sports clubs. These reactions come mainly from the lower part of the middle class, which sees in Belgium an opportunity to access 'free private schools' with higher levels of expectation and additional classes such as religious and citizenship education. Contrary to this, parents with higher degrees and intellectual occupations prefer to send their children to French schools, fearing that they would be unable to succeed at French universities later on without an education according to a French school curriculum.

Urban sociology has demonstrated that outlying suburbs and rural areas as well as old city centres are common choices for middle class homes. What is new here, and this is something that lifestyle migration studies can help us understand, is that this classical residential upgrading mobility occurs in an international context. By analyzing housing choices with respect to questions of middle class habitus, it becomes clear that Belgium represents a potential for upward social mobility. The professional position of the migrants in question has improved in comparison to the position of their parents. Belgium, instead of being a place for further career progress (as when French people move to foreign capital cities), is the place for residential upgrading. This residential strategy is strongly articulated in concerns about education and a secure environment for children. Yet, as an international mobility, CBRM involves different symbolic and material aspects than regular residential movements.

Symbolic and material aspects of CBRM

Different social classes do not have the same access to international mobility, and national borders matter less for the most privileged. As Wagner writes, 'the reality of borders has almost always been a matter of class' (2007, p.43). This means that the resources that allow people to move abroad, to enter international institutions or economic and cultural clubs, to learn foreign languages, or in the case of Lille, to manage a short-distance expatriation from an administrative and legal point of view, are not equally distributed. Looking at the middle class, Wagner argues that it has a special relation to travel and tourism (Wagner, 2007, p.87). Foreign countries are appealing because they offer exoticism and a promise of 'authenticity'. They also offer a chance to reinvent oneself and to put national social determinants at a distance.

With CBRM, this appeal takes on other features that are both symbolic and material: it reinforces the decision-making process, offers a reasonable share of adventure and requires the acknowledgement of new legal norms. First, the factor of internationality of CBRM is observable in the way migrants depict how they made their residential choice. The national border puts at a distance the constraints that were encountered in the previous national territory and opens a new realm, symbolic of a fresh start that nevertheless does not break the links with France. Considering this, the possibility of settling in Belgium opens households' residential horizons and improves their power of arbitration. It gives credit to their ability to articulate personal ambition and smart decisions. Therefore, in addition to the material improvement of housing conditions and the everyday life changes that occur as a consequence of CBRM, migration reasserts the decision-making power of the middle classes. As it is depicted in the following interview excerpts, the international context gives migrants the opportunity to develop real strategies that include financial, material and social aspects.

> We had already visited several houses in the street because we had already started to prospect, thinking 'Well, we are currently renting but we want to buy soon' […] And we thought … we were told that it was more interesting to live in Belgium for civil servants regarding taxes. So we looked for more information; we downloaded forty pages of documents … I mean we read a lot, we did not decide just like that. […] And there was also the financial aspect of real estate being a lot cheaper than in France. For this house, on the other side of the border, we would have paid twice the price. 'Why Belgium?' I always told B [her partner], if one day I have to leave the north [of France] what I will miss most is Belgium. In fact my grandmother was born in Messines in Belgium […] At my grandmother's we used to speak Flemish; I have Flemish roots.

Mrs C., a 37-year-old teacher, explains that the choice she made with her partner to move to Belgium was a thoroughly thought-out decision that combined fiscal, residential and affective aspects. This step is also a way to commit in a strong family project, as she states later: 'We did not want to buy a temporary house, we wanted a house where we could spend our life.'

Another aspect of the cross-border context of residential mobility is that it offers an opportunity to experience life abroad without losing one's points of reference and feeling insecure. Therefore, interviews not only reveal residential strategies, but also the representation that migrants have of themselves as being able to live in a foreign yet familiar country. In the excerpt below a 50-year-old woman, who is a junior executive and rents a house in Tournai, hoping to buy a house there, explains the way she sees life in Belgium.

> **Mrs O.:** What I mean is that you can have a quality of life here … If you question yourself a bit, if you are a bit adventurous … That's it, you need to get off the beaten track, that is forced on you, that society imposes. All this for what? To fatten the banks? In the end that's what we do.
> **Me:** So being in Belgium for you is something that is original?
> **Mrs O.:** Yes, for sure. Without a doubt. And I am proud of it. I'm happy, you know, that I had this courage, because nobody does it because it's scary, since you arrive in a new country. It's close indeed, but still we're not home here. You can feel this from the French part.

She depicts her move to Belgium as an adventure that she can live because she learnt to be open-minded and to 'get off the beaten track'. She underlines her bravery as a way to distinguish herself from the rest of her milieu. The international experience becomes a way to obtain a new social status. Migrants can picture themselves as pioneers in exotic territories even though they are aware that Belgium is a very familiar country for French residents. This attitude is especially pronounced among the most affluent migrants in the sample.

Finally, for those with 'little means', the international dimension of CBRM is expressed more in terms of the constraints and difficulties of adapting to new legal rules. They have less experience of being abroad and sometimes have insufficient social and cultural resources to master fiscal or administrative issues. They sometimes face situations in which their financial stability is put at risk by the migration. In some cases, they envisage a return to France. They also feel a greater distance between their daily life and the process of European construction. Before experiencing migration, they expected Europe to be harmonized and viewed the Schengen area as a territory where anyone could settle without hindrance.

> And I tell myself 'But where is Europe?' […] I thought it was going to be much more unified, in terms of legal rights and functioning. First, I had a shock about work, unemployment benefit and so on … It has nothing to do with France. It is not the same interlocutors.
>
> (Mrs. C., 30 years old, employee in the audio-visual sector, rents a house in Tournai with her partner, an artist)

During the migration process they find out that they have to comply with the national rules of Belgium and this can generate resentment vis-à-vis the European Union and its representatives, as expressed by this 50-year-old woman who is unemployed and owns a house with her husband in Mouscron.

> Deep down I feel mad! Very mad about the system that was imposed on us, that was put in place. This Europe thing is chaotic!

In contrast to this, more educated migrants use their cultural capital to reinforce their international position. One example of this is given by a junior executive who plans to write a book about how to settle in Belgium and all the legal and cultural aspects that the move entails. He is able to transfer his capital at an international level and gains credit and legitimacy from it.

These elements suggest that CBRM is not only about dealing with 'being abroad' and 'nearby home' at the same time, but also about how different forms of capital – economic, social, cultural and also international – are mobilized in order to improve a residential and social position. The economic and cultural capital of the middle class is converted into symbolic capital when migrants access higher residential positions or receive social recognition for being able to settle abroad. However, the internationality linked to this form of residential move can also become challenging for those who move when it becomes difficult to deal with the legal and material aspects of migration.

Conclusion

Facing a tough housing market, French households are looking for cheaper houses in the neighbouring cities, which are also attractive for their architectural characteristics, their accessibility and the change in fiscal treatment that settling in Belgium offers. Expatriation to cities like Mouscron is thus a response to the appeal of cheaper housing and better quality of life. Estate agencies have understood this attraction of the border market to French households, and over recent years they developed a series of commercial devices specifically designed for them, like a list of the advantages and drawbacks of settling in Belgium that includes elements about taxation, health, legal rights and more. However, this chapter makes the case for a different way of looking at CBRM, which pays more attention to the social determinants of migration and the distinction processes it encompasses. It shows that even among members of the middle class, variations exist in the strategies that are formulated by households, and that these variations can also be found regarding the importance given to the international dimension of CBRM.

The migrants' choices are driven by their social belonging and the socialization processes they encountered. The data gathered in this research show that CBRM is linked to a strong 'border legacy' that brings about specific dispositions. Moreover, migrations of middle class individuals can be described as a reaction to what Eric Maurin (2004) calls a 'fear of relegation', which consists in making residential projects to maintain or improve a social position: avoiding social housing and depreciated neighbourhoods, offering the best housing conditions for children, and coupling residential and educational strategies to improve the social conditions of the next generation. Here, it is dwelling, rather than a professional career, that bears the responsibility of upgrading a family's or an individual's situation.

In the case of CBRM, the idea of migration as a 'fresh start' or a 'new beginning' (Benson and Osbaldiston, 2014, p.218) seems to be less present than in what is described as lifestyle migration, due to the fact that migrants preserve most of the links they had with France. There is no real rupture between previous and new lives. However, the fact that residential mobility takes place in an international context allows for a stronger redefinition of social position and identity than in the case of national residential mobility. It generates a feeling of 'exoticism' and brings about the need for new arrangements and strategies in daily life to adapt to the new residential position. CBRM does not provide a way to live the 'antithesis of life before migration' (Benson and O'Reilly, 2009, p.611), as might be the case in long-distance migrations. Distance does matter and spatial proximity makes international residential migration less of a radical change. Yet the logics identified in the context of lifestyle migration regarding the process and the representations of a new lifestyle offered by migration are still valid.

While European societies are overwhelmed with liberal ideas that encourage mobility and flexibility, it seems important to remember that international capital is not equally distributed among people and that border crossing remains a matter of class. Whereas CBRM is often depicted as a self-evident phenomenon or a natural consequence of European integration, it also proves discriminating. Affluent citizens can use their resources to secure their position abroad and gain international experience that is transformed into social distinction. The lower contingent of middle class migrants, however, faces difficulties that remind us that nations still have control over migration policies, and that it requires various types of capital to understand and benefit from them. At a time when migration remains a highly sensitive matter and the Schengen area is seriously called into question by some governments, it seems important to remember that even at so-called open borders, there are ongoing processes of social selection and distinctiveness.

Notes

1 In 1886 there were 500,000 Belgians in France, mainly settled in the Nord department. In 1870 they represented 55 per cent of Roubaix's inhabitants (Rainhorn, 2008; Noiriel, 1988).

2 This research uses data from interviews (n=30) carried out with inhabitants who have experienced cross-border residential mobility and also with some who have not and remained in France. It offers a comparison point and helps extract what is specific about cross-border residential trajectories. However, a significant majority of the interviews

were conducted with households that left the Lille agglomeration to settle in one of the border cities while the individuals maintained their jobs in France. The study population has common features in terms of job status, income and housing conditions. Fieldwork also demonstrates that they have similar social backgrounds that may account for their choice to migrate.

3 *Insee, RP2007 et RP2012.*

4 All excerpts are my translations of the original French.

5 The interviewed population also includes people renting apartments with the prospect of later becoming owners.

Bibliography

ADULM, 2010. *Les marchés du logement dans les territoires belges de l'Aire Métropolitaine de Lille.*

ADULM, 2012. *Diagnostic du 2e programme local pour l'habitat de Lille Métropole 2012/2018.*

Balogh, P., 2013. Sleeping abroad but working at home: Cross-border residential mobility between transnationalism and (re)bordering. *Geografiska Annaler: Series B, Human Geography*, 95(2), pp.189–204.

Bauer, P. and Darley, M., eds., 2007. *Borders of the European Union: Strategies of crossing and resistance*. CEFRES.

Benson, M. and Osbaldiston N., 2014. *Understanding lifestyle migration. Theoretical approaches to migration and quest for a better way of life.* New York: Palgrave Macmillan.

Benson, M., 2013. Living the "Real" Dream in *la France profonde*? Lifestyle Migration, Social Distinction, and the Authenticicities of Everyday Life. *Anthropological Quaterly*, 86(2), pp.501–526.

Benson, M. and O'Reilly, O., 2009. Migration and the search for a better way of life: A critical exploration of lifestyle migration. *The Sociological Review*, 57(4), pp.608–625.

Bidou, C., 1984. *Les aventuriers du quotidien. Essai sur les nouvelles classes moyennes.* Paris: PUF.

Boesen, E., Schnuer, G. and Wille, Ch., 2015. Urbanity in Rural Space. Residential Migration in the German-Luxembourgish Border Region. In R. Garstenauer and A. Unterwurzacher, eds., 2015. *Aufbrechen, Arbeiten, Ankommen. Mobilität und Migration im ländlichen Raum seit 1945*. Innsbruck: Studienverlag, pp.225–244.

Bolzman, C. and Vial, M., 2007. *Migrants au quotidien: Les frontaliers. Pratiques, représentations et identités collectives*. Zurich, Geneva: Seismo.

Bourdieu, P., 1979. *La distinction: critique sociale du jugement.* Paris: Les Editions de Minuit.

Carpentier, S., 2012. Cross-border local mobility between Luxembourg and the Walloon Region: An overview. *European Journal of Transport and infrastructure Research*, 12 (2), pp.198–210.

Cartier, M., Coutant, I., Masclet, O. and Siblot, Y., 2008. *La France des "petits moyens". Enquêtes sur la banlieue pavillonnaire.* Paris: La Découverte.

Charmes, E., 2005. *La vie périurbaine face à la menace des gated communities.* Paris: L'Harmattan.

Clerval, A., 2013. *Paris sans le peuple. La gentrification de la capitale.* Paris: La Découverte.

Collet, A., 2012a. Montreuil, "le 21e arrondissement de Paris"? La gentrification ou la fabrication d'un quartier ancien de centre-ville. *Actes de la recherche en sciences sociales*, 195, pp.12–37.

Collet, A., 2012b. Le loft: habitat atypique et innovation sociale pour deux générations de "nouvelles classes moyennes". *Espaces et sociétés*, 2012(1), pp.37–52.

Cusin, F., 2012. Le logement, facteur de sécurisation pour des classes moyennes fragilisées? *Espaces et sociétés*, 2012(1), pp.17–36.

Darmon, M., 2010. *La socialisation.* Paris: Armand Colin.

Dietrich-Ragon, P., 2013. Classement, déclassement, reclassement sur le marché residential. L'exemple des occupants de logements dégradés parisiens. *Revue française de sociologie*, 54, pp.369–400.

Donnan, H. and Wilson, T., 1999. *Borders: Frontiers of Identity, Nation and State*. Oxford: Berg.

Gellereau, C., 2014. L'apport du terrain et de la ville dans la définition d'un objet de recherche : le cas de la catégorisation des 'expatrié'. *Communication pour l'Atelier Doctoral du Réseau Thématique n°9 de l'Association Française de Sociologie.* Lille 1 University, Unpublished.

Grafmeyer, Y. and Authier, J.-Y., 2008. *Sociologie urbaine.* Paris: Armand Colin.

Gielis, R., 2009. Borders make the difference: migrant transnationalism as a border experience. *Tijdschrift voor Economische en Sociale Geografie*, 100(5), pp.598–609

Hamez, G., 2004. *Du transfrontalier au transnational : approche géographique. L'exemple de la frontière franco-belge*, Thèse de doctorat en géographie à l'Université de Paris I.

Haumont, N., 1966. *Les pavillonnaires. Etudes psycho-sociologique d'un mode d'habitat.* Paris: Centre de Recherche d'Urbanisme et Institut de Sociologie urbaine.

Jagodic, D., 2014. Cross-border residential mobility in the Italian-Slovenian borderland: a transnational laboratory for the creation of European space. In M. Bufon, J. Minghi and A. Paasi, eds., 2014. *The New European Frontiers: Social and Spatial (Re) Integration Issues in Multicultural and Border Regions.* Newcastle: Cambridge Scholars Publishing, pp.173–196.

Le Galès, P., 2005. Interesting times for urban sociology. In T. May and B. Perry, eds., 2005. The Future of Urban Sociology. *Sociology*, 39(2), pp.343–370.

Maurin, E., 2009. *La peur du déclassement. Une sociologie des récessions.* Paris: Le Seuil.

Miot, Y., 2012. La ségrégation socio-spatiale dans la métropole lilloise et à Roubaix: l'apport des mobilités résidentielles. *Géographie, économie, société*, 14(2), pp.171–195.

Noiriel, G., 1988. *Le creuset français, Histoire de l'immigration (XIXe siècle-XXe siècle).* Paris: Le Seuil.

Oberti, M. and Préteceille, E., 2004. Les classes moyennes et la ségrégation urbaine. *Education et sociétés*, 14(2), pp.135–153.

Oliver, C. and O'Reilly, K., 2010. A Bourdieusian Analysis of Class and Migration: Habitus and the Individualizing Process. *Sociology*, 44(1), pp.49–66.

Paris, D., 2002. Lille, de la metropole à la region urbaine. *Mappemonde*, 66(2), pp.1–7

Portes, A., 1999. La mondialisation par le bas. *Actes de la recherche en sciences sociales*, 129, pp.15–25.

Rainhorn, J., 2008. Le Nord-Pas de Calais, une région frontalière au cœur de l'Europe. *Hommes et migrations*, 1273, pp.18–34.

Rérat, P., Moine, A., Gertsch, K. and Signoret, P., 2011. La mobilité résidentielle transfrontalière dans l'Arc jurassien franco-suisse. *Géo-Regards, Revue Neuchâteloise de géographie*, 4, pp.153–168.

Sassen, S., 2012. *Cities in a World Economy*. London: Sage Publications.
SPF Économie, 2011. *Population par nationalité et sexe; superficie et densité de la population (2008)*. Belgium.
Van Zanten, A., 2009. *Choisir son école. Stratégies familiales et médiations locales*. Paris: Puf, Le lien social.
Wagner, A.-C., 2007. *Les classes sociales dans la mondialisation*. Paris: La Découverte.

11 Asymmetries in the formation of the transnational borderland in the Slovak–Hungarian border region

Tamás Hardi

Introduction

A significant social experience of the present day is the increasing permeability of the internal borders of the European Union, a result of the elimination of the physical barriers to border crossing, allowing freedom of travel. It is difficult to overstate the importance of this change for those who, years or decades ago, were not permitted to cross the border, and also for those who had to wait for hours, often in almost intolerable conditions, at border crossing stations and were only able to continue their travels after humiliating bureaucratic procedures. In most cases the partial and then the complete opening of the borders has been welcomed with enthusiasm, and the ever greater permeability of the borders serves the integration of the economic and political macro-systems of nation states, leading to the emergence of an enlarging supranational economic and political space in which the free flow of persons, labour, goods and capital is secured. In this system border regions occupy a unique place and, it can be argued, play a unique role. Will this cross-border flow pass them by, will they remain struggling peripheries of the nation states, or can they use the opportunities offered by this new situation? Is space opening up for them too? And what role do they play in the process of opening up?

Borders and border areas are all unique, individual contexts. The birth, change and character of the spatial borders depend to a large extent on the spatial unit (in this case the state) that they surround, but this is a symbiotic relationship: states, border regions and the characteristics of the state border all influence each other (Hardi, 2001). This complexity must be taken into account in our research on border regions. It is not enough to analyse the political, social, economic and cultural features of these regions by the comparison of case studies; we have to strive for a deeper understanding by combining historical contextualization with a general theoretical framework.

In the following we will try to illustrate such an approach by drawing on the study of the Slovak–Hungarian border. The concrete research question is in what circumstances and to what extent the population of the Slovak–Hungarian border regions can make use of the possibilities offered by the accessibility of the 'other side', and if there is a chance for the birth of single cross-border regions and a transnational society.

The paper is built on the findings of three questionnaire surveys:

- In 2008 a survey was conducted along the total length of the Slovak–Hungarian border region, with layered sampling, questioning 1,000 people on each side of the border. The research area was a border area defined by using a functional approach (Hardi and Tóth, 2008).[1]
- In 2010 we examined the situation of those living in suburban Bratislava, comparing the situation of the population living on the Slovak side and the population that moved to the Hungarian side. Using the snowball method we examined the group of those moving out after 2004; 360 inhabitants were questioned on the Slovak side and 240 on the Hungarian side (Hardi, Lados and Tóth, 2010).[2]
- In 2014 we conducted a mental map survey among students; in Slovakia 194 students of Slovak nationality and 69 students of Hungarian nationality were interrogated in five cities distributed evenly across the country, while in Hungary, 244 students (all Hungarian nationals) were asked in seven cities (Hardi, 2015).[3]

The applicability of Martinez's borderland theory to the case of Hungary and its neighbours

A dynamic definition of borderlands was formulated by Martinez, who defines four types of border areas on the basis of the number and depth of cross-border interactions. In 'alienated borderlands' tension prevails; the border is functionally closed, and cross-border interaction is totally or nearly totally absent while the residents of each country act as strangers to each other. In 'co-existent borderlands' stability is an on/off proposition; the border remains slightly open, allowing for the development of limited binational interaction – the residents of each country deal with each other as casual acquaintances, but borderlanders develop closer relationships. In 'interdependent borderlands' stability prevails most of the time; economic and social complementarity prompt increased cross-border interaction, leading to expansion of borderlands – borderlanders carry on friendly and cooperative relationships. Finally, in 'integrated borderlands' stability is strong and permanent; the economies of both countries are functionally merged and there is unrestricted movement of people and goods across the boundary – borderlanders perceive themselves as members of one social system (Martinez, 1994, p.3). This typology implies the notion of change in the sense of a gradual increase of transborder interaction and transnationalism.

We believe that the borderland types defined by Martinez are perfectly applicable to the border regions of the former socialist countries and help us to understand the changes that they underwent over the last quarter of a century. In the period before the regime change, the modes of interaction that characterized the respective state borders were not the same as afterwards. Although totally closed borders and alienated border regions arguably only existed during the early 1950s, after which borders became somewhat more permeable, it is undeniable that daily or regular border crossing was only possible for a narrow layer of the population. However, it is important to consider individual cases and particularities. While on the Soviet-Hungarian and the Austrian-Hungarian border permeability was severely restricted, the Slovak–Hungarian relationship was more liberal from the 1970s. In fact, all the Hungarian borders can be described as having been of the 'co-existent' type in the 1970s and 1980s, but at certain sections the centralised state approved of specific alleviations, allowing regular border crossing of the population of the border region, including commutes to work in some cases. At the time of the regime change the border regions of Hungary gradually turned into interdependent borderlands; one of these is the border region treated in our paper. After the accession of Slovakia and Hungary to the European Union we have seen some sections of the borders develop into integrated borderlands, but so far this has happened in just a few places and only to a limited extent.

Another important element in Martinez's theory is the introduction of the notion of borderland society. In this society we can distinguish so-called national borderlanders and transnational borderlanders, whose proportion correlates to the integration level of the border region. While in the case of alienated borderlands there is practically no person who belongs to the latter group, in integrated borderlands their proportion is high.

But what defines these two types? In the words of Martinez:

> National borderlanders are people who, while subject to foreign economic and cultural influences, have low-level or superficial contact with the opposite side of the border owing to their indifference to their next-door neighbours or their unwillingness or inability to function in any substantive way in another society. Transnational borderlanders, by contrast, are individuals who maintain significant ties with the neighbouring nation; they seek to overcome obstacles that impede such contact and they take advantage of every opportunity to visit, shop, work, study, or live intermittently on the 'other side'.
>
> (1994, p.6)

Joining the Schengen Agreement and the resulting opening of borders allowed the development of increasingly denser and more in-depth interactions in the border regions of Hungary. The everyday experiences of the author (confirmed by the results of the survey) show, however, that characteristic transnational borderlanders have appeared in very narrow places and in very limited numbers, and their distribution is asymmetric (this transformation is more typical of the inhabitants on the Slovak side than those on the Hungarian side of the common border). An important feature of the last twenty years in Eastern Europe is the fact that, parallel to the weakening of the separating role of borders, nationalism has become palpably stronger, and the opening borders also reinforced the sense of danger represented by the other side. This evidently slows down the integration process as it blocks, among other things, the development of cross-border infrastructure (the construction of new border crossing stations, roads, bridges, etc.).

The Slovak–Hungarian border region is interesting for our topic because among all borders of Hungary this one contains areas that come closest to the borderland category characterized as integrated by Martinez, meaning that they have the highest number of people who 'use' the other side of the border on a daily basis, and according to our survey the proportion of transnational borderlanders is significant here. We have to admit, though, that there are substantial differences between the western and eastern sections of this border: advanced integration is mainly present in the western section of the Slovak–Hungarian border and even there it is restricted to the hinterlands of major cities; also, interactions in significant numbers and depth are primarily typical for the Slovak side.

We think it is important to remark that the theory of Martinez reflects a concept of nation, which builds on the category of political nation. In the region under study here, however, concepts of nation and nationalism are stronger, which integrate the groups living outside the state territory as parts of the nation, but belonging to the cultural/linguistic nation as well (e.g. the Hungarian ethnic minority that lives in Slovakia). These minorities see themselves as part of a culturally defined mother nation and typically identify less with their political nation. As the chapter will reveal, the group of transnational borderlanders mainly emerges out of these minorities, so the 'transnational' attribute has a special interpretation in this region. The members of this group are not made 'transnational' by commuting to the other side of the border and adopting the language and culture of the other nation, but by the fact that they feel that they belong there (too), while the language and culture of the nation state they live in is something that they have adopted (whether at school, in families of mixed ethnicity or in the residential environment).

In addition, our research demonstrates a higher cross-border activity by the Slovak ethnic population on the Slovak side than by the Hungarians living in Hungary. We can thus distinguish a specific group of transnationals who – especially in the region of Bratislava – establish a cross-border living space centred on the urban hinterland and driven by economic rationality, and whose transborder way of life is not motivated by ethnic or language reasons but by the practicality of everyday life.

To summarize, we can describe the Slovak–Hungarian border as giving rise to a multiplicity of borderlands. On the one hand there is a clear west-east difference in the degree of transborder contact, which is linked above all to geographical and economic development reasons. On the other hand, we find asymmetry between the two sides of the border, namely a higher degree of transborder activity on the Slovak side.

An introduction to the border and the border region

Historical and geographical features of the border

The Slovak–Hungarian border region is situated in an area that belonged to single-state formations until the end of the First World War: the Kingdom of Hungary and the Austro-Hungarian Monarchy. After the foundation of the Hungarian state (in 1000) the territory of the present Slovakia was part of the Kingdom of Hungary, which lasted, with varying levels of sovereignty, until the end of the First World War, in relatively constant geographical frameworks except for the time of the Ottoman occupation. The Slovak and the Hungarian ethnic group, despite the basically different languages, showed similar cultural features in many respects as a result of their long co-existence, and mutual assimilation was strong. The main grievance of the Slovak ethnic group is that their nineteenth-century nation building movements were confronted with strong language nationalism by the Hungarian state, and the language act passed at the end of the century tried to assimilate the Slovaks (and all other ethnicities). Of the population of the Kingdom of Hungary at the end of the nineteenth century, approximately half were Hungarians and the other half belonged to different ethnic groups.

The ethnic relations in the whole of the Austro-Hungarian monarchy were just as (or even more) complex. The centralizing language nationalism, the secession efforts of the strengthening ethnic minorities and the geopolitical efforts of the super powers together led to the disintegration of the Kingdom of Hungary and the monarchy. In the newly created Czechoslovakia language nationalism was just as prevalent as in the former state formations: after the Second World War masses of Hungarians were expatriated for the sake of ethnic homogenisation, and in independent Slovakia, formed after 1993, the ethnic and language intolerance against the Hungarians has often been visible as well.

The state border between Czechoslovakia and Hungary was created by the peace treaty concluding the First World War. This border did not follow the ethnic border, as significant areas inhabited by a Hungarian majority were annexed to Czechoslovakia. In the peace treaty of 1920 the border was designated mainly on the basis of economic, military-strategic and transport geographical considerations (Hevesi and Kocsis, 2003), and so personal relations, settlement networks and ethnic considerations, all important for the organisation of the everyday life, were not considered during the decision-making. The border was pushed northwards in connection with the Munich Agreement in 1938, and then in 1939 Hungary occupied the Zakarpatska region (Carpatho-Ukraine) that used to belong to Czechoslovakia/Slovakia[4] at that time and belongs to the Ukraine now, which also modified the contemporary Slovak–Hungarian state border (Map 11.1) and resulted in the return to Hungary of south Slovak areas inhabited by Hungarian majorities. This situation existed until the end of the Second World War.

The ceasefire agreement (1945) restored the situation that existed before 1938, and the peace treaty annexed another small area (part of the Bratislava region) to Czechoslovakia. After the disintegration of Czechoslovakia, Slovakia inherited the state borders.

In the Slovak historical consciousness the interpretation of history tries to relativize the Hungarian supremacy, and uses the expression 'Uhorsko' to denote the historical multi-ethnic Hungarian state, while the expression 'Madarsko' is applied to specify the area inhabited by ethnic Hungarians and also the modern Hungary. According to this view, in the common historical state the ethnic

Map 11.1 The Slovak–Hungarian borderland and the state border modification in 1938–1939.
Source: The author.

Hungarians oppressed other nationalities, forcing these groups to secede from the state and re-establish their sovereign national existence that had been interrupted in the ninth century. On the other hand, the Hungarian interpretation links the historical Hungary to the existence of the Hungarian state, and considers the territory of Slovakia (together with the territories annexed from Hungary after the Second World War) as lost parts of the country.

The Slovak–Hungarian state border is 679 kilometres long, the longest border section in both Slovakia and Hungary. On the western part of the border it is formed by the Danube River for approximately 143 kilometres. Crossing this section is assisted by three bridges and a tourism-oriented ferry. In this area we find significant cities and city pairs in the proximity of the border: Bratislava, Győr, Komárom/Komarno and Esztergom/Sturovo. The middle and eastern sections of the border are a mountainous area, partly riparian (the Ipoly/Ipel River comprises the border at a length of 143 kilometres) and partly land-based. This area has smaller, more peripheral towns. In the easternmost section we find large cities again, although not in the direct proximity of the border, and a landscape of gentle hills and plains. On the basis of geographical and urban network features, the border region can be divided into five sections.

1 The agglomeration of Bratislava. This involves the traditional suburban zone of the Slovakian side up to Somorja, the main commuting region of the capital city of Slovakia. The agglomeration of Bratislava has now reached across the state border and involves the area of Mosonmagyaróvár close to the border, and also some Austrian territories. The agglomeration is contiguous to the agglomeration of Vienna, and the impact of the two capital cities is jointly shaping the area.

2 The zone of the Danube cities. This entails Győr and the so-called Danube city pairs, the most prominent of which are Komarno/Komárom and Sturovo/Esztergom. It is especially the geographical location of the two Komárom settlements and Győr that led to the birth of considerable cross-border catchment areas. The special importance of these city pairs is indicated by the fact that they can now be taken as single urban agglomerations. Together they have a population in excess of 50,000, so their common services and economic attraction are equal to that of a medium-sized Hungarian city, not to mention the high density of population in the economic agglomeration along the right bank of the Danube River (from Almásfüzitő to Dorog).

3 The zone of the mountainous towns. This zone reaches from the mouth of the Ipoly River to the edge of the hinterlands of Košice and Miskolc. Its western part is adjacent to the agglomeration of Budapest, including Vác. On the other hand, the low level of urbanisation along the Ipoly River is also due to the drainage effect of Budapest. The area between the Börzsöny Mountains and the Ipoly River gravitates to the city of Esztergom, which is made possible by the Schengen borders and the existing bridges across the Ipoly and the Danube. At the northern foot of the North Hungarian Mountain Range there are the aforementioned towns with 20,000–50,000 inhabitants (Salgótarján, Ózd, Kazincbarcika, Lučenec, Rimavská Sobota), but they are somewhat further from the border (10 to 20 kilometres). Directly on the border we only find smaller centres (Šahy, Balassagyarmat).

4 The hinterlands of Košice and Miskolc. The border regions of the two cities are characterized by a deficient urban system, especially on the Hungarian side (being one of the least urbanized areas in Hungary). North of Edelény in the Zemplén Mountains we do not find any major central settlement. The areas directly on the border may gravitate to Košice more than to Miskolc, even on the Hungarian side.

5 The area of the triple border in the east. This region has a weak urban network in both countries. Smaller centres can only be found on the Hungarian side, such as Sátoraljaújhely and Sárospatak, and especially the latter has strong cross-border attraction. On the Slovakian side, Trebišov can be found a bit farther from the border, and its services are too weak to have attraction on the Hungarian side of the border as well.

The question of ethnicity and languages

In cross-border interactions, an important role is played by the ethnic and language relations of the border regions. In Slovakia the largest ethnic minority is Hungarian. According to 2011 census data, from the total population of Slovakia (5.4 million people) 8.5 per cent declared themselves as ethnic Hungarians, and 9.4 per cent stated that their mother tongue was Hungarian. The approximately 450,000 ethnic Hungarian citizens are concentrated in the southern areas of Slovakia bordering Hungary. An important feature of the border region is the strong presence of a Hungarian ethnic population in the immediate border districts of the Slovak side, reaching in some cases 80 per cent of the local population. In contrast, on the Hungarian side the proportion of the population with Slovak ethnicity and mother tongue is negligible. Hungary has a Slovak ethnic minority, but it made up approximately 0.3 per cent of the total population of 9.9 million people in 2011, and Slovaks comprise the majority in only a few settlements. What is more, those settlements that have a significant number and proportion of Slovak residents are found in the southeastern part of the country, in the vicinity of the Hungarian-Romanian border, and very few are located in the vicinity of the Slovak–Hungarian border (see Map 11.2).

This disparity also impacts language skills. Slovak and Hungarian languages are fundamentally different. While the Slovak language belongs to the family of Western Slavic languages, Hungarian is a Finno-Ugric language and its closest relatives are Estonian and Finnish. As we have shown in the demonstration of the ethnic relations, the majority of the inhabitants living on the Slovak side of the border region speak Hungarian; in fact, they consider Hungarian to be their mother tongue, and even a large portion of the ethnic Slovak population speaks or understands Hungarian, due to the significant proportion of Hungarians living in their environment. On the other hand, on the Hungarian side only a very small proportion of the population speaks or understands Slovak.

In 2008 we conducted a survey among the inhabitants of the border region concerning language skills, with 1,000 inhabitants on the Slovak and another 1,000 on the Hungarian side (Figure 11.1). In Slovakia, 68 per cent of the

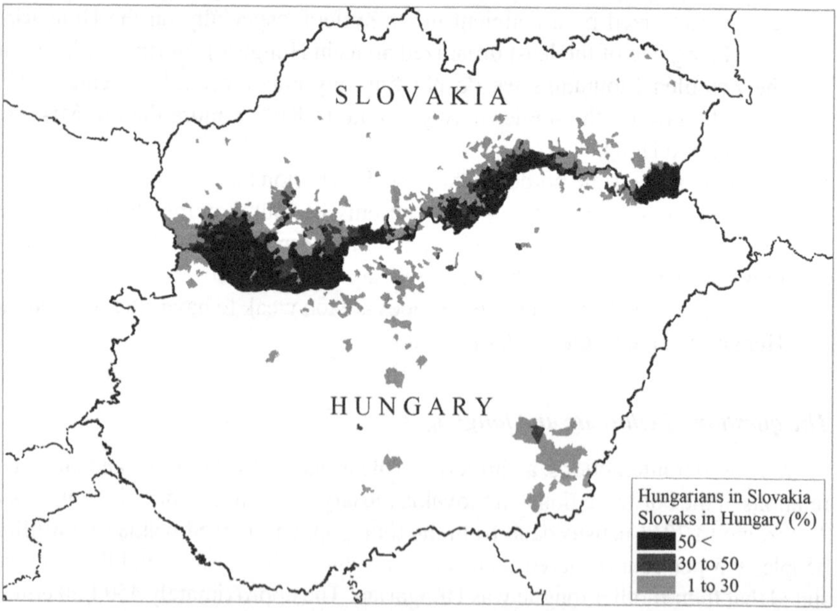

Map 11.2 The share of ethnic Hungarians in Slovakia and ethnic Slovaks in Hungary by settlements, 2011.
Source: The author, 2011 Census (Data).

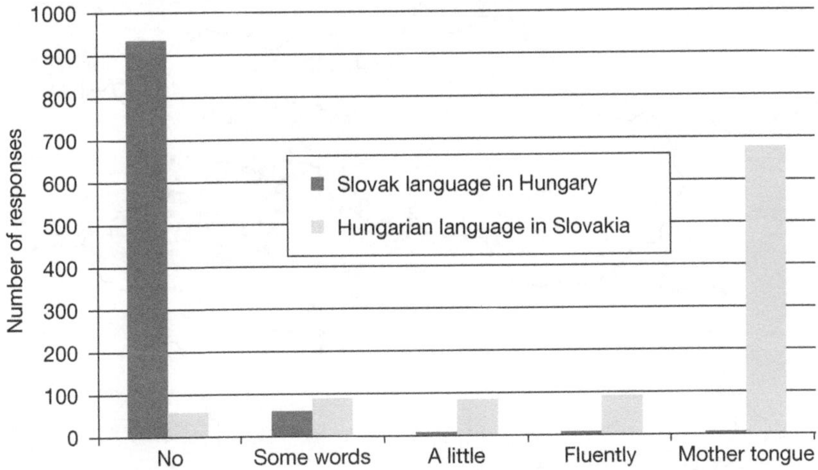

Figure 11.1 Language knowledge in the Slovak–Hungarian borderland.
Source: The author, 2008 questionnaires (Data).

respondents declared Hungarian to be their mother tongue, and the majority of the other respondents mentioned language skill at some level, while the proportion of those who said they did not speak Hungarian was only 6 per cent. On the other hand, only three people of those asked on the Hungarian side declared Slovak their mother tongue, and 93 per cent indicated that they did not know the language of the northern neighbour country.

Another important ethnic feature of the border region is the high proportion of inhabitants belonging to the Roma ethnicity. In the censuses of 2011, 2 per cent of the population in Slovakia and 3 per cent in Hungary declared that they were of Roma ethnicity. However, the real figures of the Roma population are much higher than registered in censuses, as many do not want to admit their Roma identity in the questioning session. In the eastern part of the border region, the share of Roma ethnic minority is particularly high (Map 11.3), especially in the isolated, peripheral areas along the border where their high proportion is accompanied by extreme poverty and segregation. In this respect the state border is not a separating line, as both states struggle with similar problems. Parallel to the dynamic development of the western part of the border region there are considerable social problems and crisis situations along the east sections.

Economic development and disparities

The territorial disparities of regional gross domestic product (GDP) clearly indicate that the most developed areas of both states can be found along the common border (Maps 11.4 and 11.5). The strong concentration of the economy in the western border section is undeniable. The capital cities of both countries have a significant and still growing share of the total production in their countries. In 2001 Bratislava accounted for 24.6 per cent of Slovakia's total GDP, and this

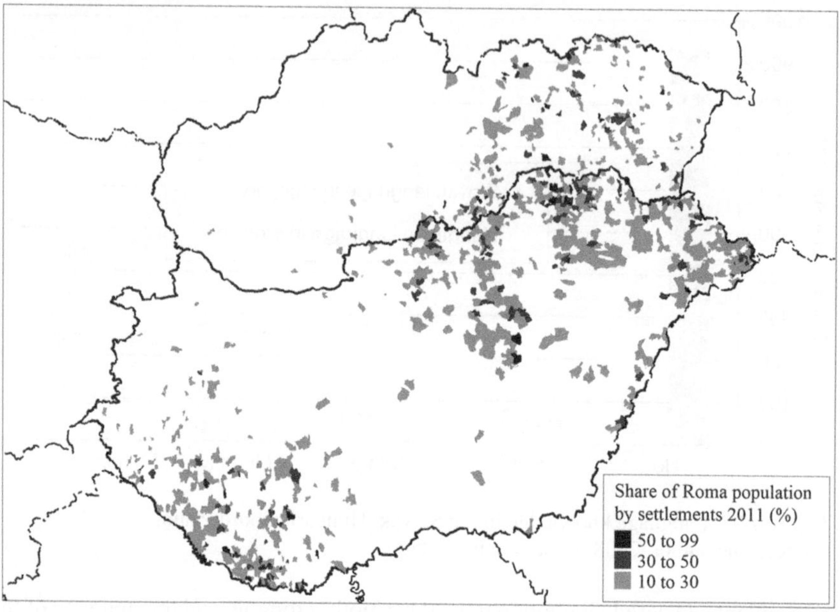

Map 11.3 The share of Roma population in Slovakia and Hungary, by settlements, 2011. Source: The author, 2011 Census (Data).

share grew to 27.6 per cent in 2011. The economic weight of Budapest is even larger, as it represented 34.7 per cent of GDP in 2001, increasing to 38.3 per cent in 2011. Among the counties along the border, the growth of the western ones is dynamic. In both Hungary and Slovakia an increase greater than the national average was typical from 2001 to 2011. In Slovakia it is Trnava that boasts of the highest production per capita after Bratislava, and during the whole of the decade these districts show the fastest growth in the country. Their share of the production of Slovakia also increased over the course of the decade, while the proportion provided by middle and eastern Slovakia decreased. In the east it is only the Košice district that shows a considerable growth, approaching the national average. All this demonstrates that the economic power of Slovakia is concentrated in the western and northern areas of the country, and in the east Košice stands out as an island. On the Hungarian side of the border it is also the western areas that show the fastest growth and the largest number of goods produced, only surpassed by the capital city. The weight of Budapest (and Pest county) within Hungary exceeds that of Bratislava and Trnava districts together in Slovakia, both in value produced and in population. Despite the basically high level of development of the western counties in Hungary, the regional disparities in the whole of Hungary and also across the border counties are bigger than in Slovakia. In Hungary the western counties (Győr-Moson-Sopron and Komárom-Esztergom) grew at a rate far above the national average in the decade in question, and have become the most developed Hungarian counties after Budapest.

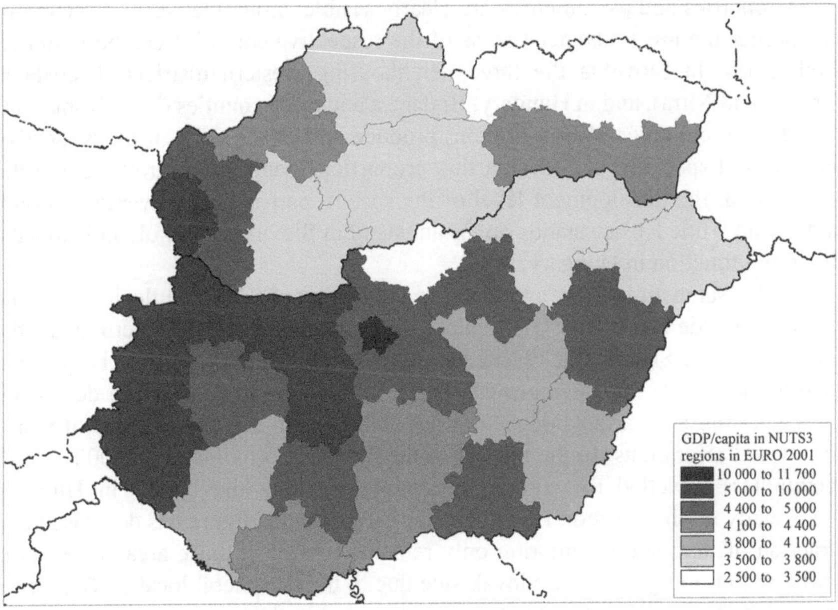

Map 11.4 Regional GDP per capita in Slovakia and Hungary at current market prices by NUTS 3 regions, 2001.
Source: The author, EUROSTAT (Data).

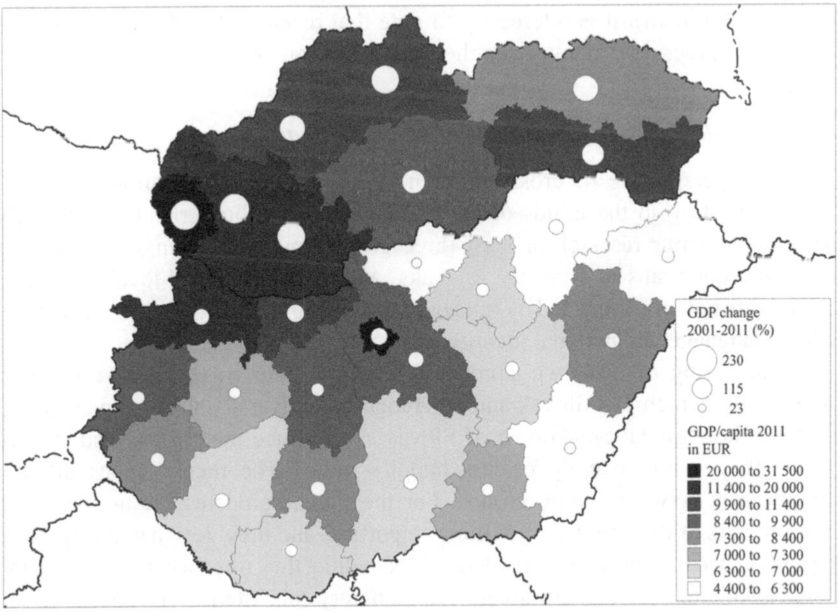

Map 11.5 Regional GDP per capita in Slovakia and Hungary at current market prices by NUTS 3 regions, 2011 and change 2011/2001 (%).
Source: The author, EUROSTAT (Data).

Symmetries and asymmetries are clearly visible: along the western section of the border the most advanced areas of the respective countries can be found on both sides. In Slovakia the three neighbouring western districts (Bratislava, Trnava and Nitra), and in Hungary, Budapest and three counties (Pest, Komárom-Esztergom and Győr-Moson-Sopron) produce half of the GDP of the respective countries. Especially in Slovakia this proportion seems to be growing. On the other hand, the development level of the eastern part is below average on both sides, and while Košice stands out as an island in Slovakia, Miskolc fails to have the same function in Hungary.

At the same time it is an interesting phenomenon that while in 2001 the Hungarian side had higher GDP values in the more advanced western areas, the districts on the Slovak side reached a higher level during the decade in question due to their more intensive growth, and on the whole they are more developed now than the Hungarian side. This is especially visible in the changes of work-motivated migrations. In the middle of the 2000s, approximately 30,000 Slovak commuters travelled to work to the facilities along the border in Hungary (especially on the western, Danubian side). To date this figure has decreased to a fraction of that, and commuting only remains typical in those areas where it is difficult to find a job on the Slovak side due to the peripheral location (e.g. from Sturovo to Esztergom). In other towns and cities this phenomenon is less common now because wages today are lower in Hungary than in Slovakia, although the difference is very small. Furthermore, the introduction of the Euro in Slovakia and the weakening of the Hungarian forint made conversion less favourable for employees than it had been during the period of an autonomous Slovak currency (the Slovak crown). It is interesting to note that recently the Slovak automotive industry has recruited labour from the Hungarian border region.

Mental borders among inhabitants of the borderland

From the perspective of cross-border interactions, the examination of mental borders existing in the minds of the population in the borderland is extremely important. In our research at least three groups must be distinguished in this respect: Hungarians in Hungary, Slovaks, and Hungarians in Slovakia. Mental borders are not only impacted by state borders, but are also influenced by historical, ethnic, language and national identity issues.

In the spring of 2014 we conducted a mental map survey in five cities in Slovakia, in which 194 Slovak and 69 Hungarian university students were asked to draw the regional breakdown of Slovakia, introduce the regions and specify where they would happily or less happily move. The mental maps differed significantly between ethnic groups. For the Slovak students the northern and eastern parts of the country were more important, and they designated respective regions in geographically correct detail. Meanwhile, they often used the expression ‘Madarsko’ for the southern areas of the country inhabited by Hungarians, and most respondents did not differentiate within this region and rejected this area as a potential destination for relocation. This southern strip of land is not a single area either geographically or historically, but it is separated from other parts of Slovakia on ethnic grounds. The Hungarian subsample, on the other hand, gave inverse answers. This difference can be traced back to the fact that the two ethnic groups have traditionally been segregated, with people of Slovak ethnicity inhabiting the mountainous regions while Hungarians lived in the plains, and so the mental maps were formed and confirmed generation after generation.

Another reason for the separation of the southern land strip is the modification of the border in 1938. This change was experienced by Hungary as a 'regaining' of the territory, and as the 'mutilation of the country' by Slovakia. As explained in the historical section, Hungary was given the southern land strip inhabited by a Hungarian majority, but after the Second World War the former borders were restored, returning the land inhabited by Hungarians to the newly formed Slovakia (see Map 11.1). This border that existed for only a few years has nevertheless left its mark on political and public attitudes. During the socialist regime it operated as an important 'crypto-border' for development policy, in so far as several economic and infrastructure developments were located north of this borderline, and not south. It is also worth mentioning that during the designation of the administrative units in the 1990s, it was also an important aspect (and a topic of debate) that the districts (*kraj*, i.e. NUTS3) should preferably have a north to south extension, so these administrative units split the homogeneous ethnic Hungarian zone. As a result of this, the ethnic characteristics and the gradual economic and infrastructure underdevelopment of the southern land strip are not visible in the territorial statistics, although the economic, modernizing and demographic centre of the country has moved northwards in recent decades.

All in all, we can say that the skills and experiences concerning the other side of the border are not the same among the groups in question. With a view to the fact that Slovakia is home to a large number of inhabitants of Hungarian ethnicity, the Slovaks, especially the ones living in the borderland, have daily contact with the Hungarians living in Slovakia; they speak or understand Hungarian and may have family members of Hungarian ethnicity. Irrespective of how they judge the Hungarians, this leads to a higher level of information and relationships. Hungarians living in Hungary have relatively little direct experience of the Slovak population. They rarely visit Slovakia, and their direct contacts are usually limited to the Hungarians living there, and so the knowledge of the Slovak language is evidently less important. Similarly limited is the experience that Slovak citizens living farther from the Hungarian inhabited areas have, and their relationship networks are also narrower, while the Hungarians living in the border region in Slovakia have the largest relationship network.

Asymmetries in cross-border interactions

Asymmetries in travel habits

The travel habits of the inhabitants living in the border area shed light on the integration level of the border region. We looked at how often and for what purposes inhabitants visited the other side of the border, and our 2008 survey allows for a more detailed analysis of the travel habits of the borderland population. At that time, 97 per cent of the interrogated ethnic Hungarians living in Slovakia had been to the other side of the border, while the figure for the inhabitants of Slovak ethnicity was 94 per cent. Only 79.9 per cent of Hungarian respondents living on the Hungarian side had been to Slovakia.

There are considerable differences in the frequency of travel as well. A significant proportion of the inhabitants on the Slovak side, approximately 28.3 per cent, travel daily, weekly or monthly to Hungary, while only 16 per cent of those living on the Hungarian side visit Slovakia regularly (Figure 11.2).

Among the respondents from Slovakia, the frequency of daily visits to the other side was somewhat higher than the average in the western part of the area, amounting to almost 5 per cent. Visits of daily and weekly frequency were usually linked to shopping and the use of services, while the main reason for irregular travels was tourism. It is clear that if more than a quarter of the population living in the borderland on the Slovak side visits Hungary daily or weekly, the number of interactions is high. The proportions have probably changed in the last years: while commuting with employment motivation is less typical now, using the other side of the border for business purposes has become more frequent, as has the moving of Slovak citizens to Hungary for residential purposes. In contrast, the number of border crossings from Slovakia to Hungary for shopping is now higher than in the opposite direction. Due to the introduction of the Euro in Slovakia and the worsening exchange rates of the Hungarian currency, it is economically beneficial only for the citizens on the Slovak side to use Hungarian shops and services, whereas crossing for shopping in the opposite direction has practically ceased. The asymmetry in this respect has presumably grown since 2008.

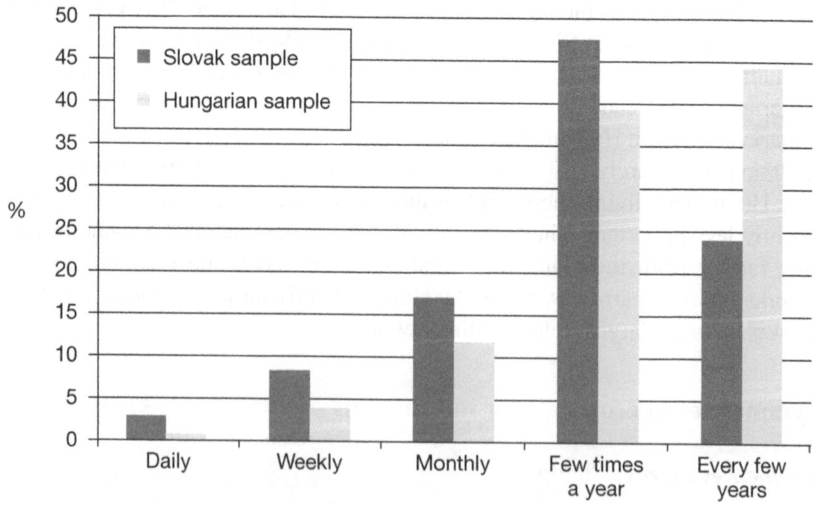

Figure 11.2 Frequency of visiting the other side.
Source: The author, 2008 questionnaires (Data).

Asymmetries in personal relations

We also looked at the personal relations network of the respondents in 2008, which also reflected visible differences between the two sides of the border. Under the category 'personal relations' we included family and friends as well as workplace and business relationships. Among the respondents on the Slovak side, 53.1 per cent had some personal contacts on the other side of the border, while the figure for those on the Hungarian side was only 25.7 per cent (Figure 11.3). Ethnic differences could of course be seen on the Slovak side. While 62.3 per cent of the respondents with Hungarian nationality maintained some personal relations on the other side, only 34.5 per cent of the Slovaks did so.

If we look at the complexity of relationships, i.e. the diversity of the relationships maintained by one respondent, we again find higher degrees on the Slovak side (Figure 11.4). A typical respondent on the Slovak side has significantly more types of relationships on the Hungarian side than the other way round.

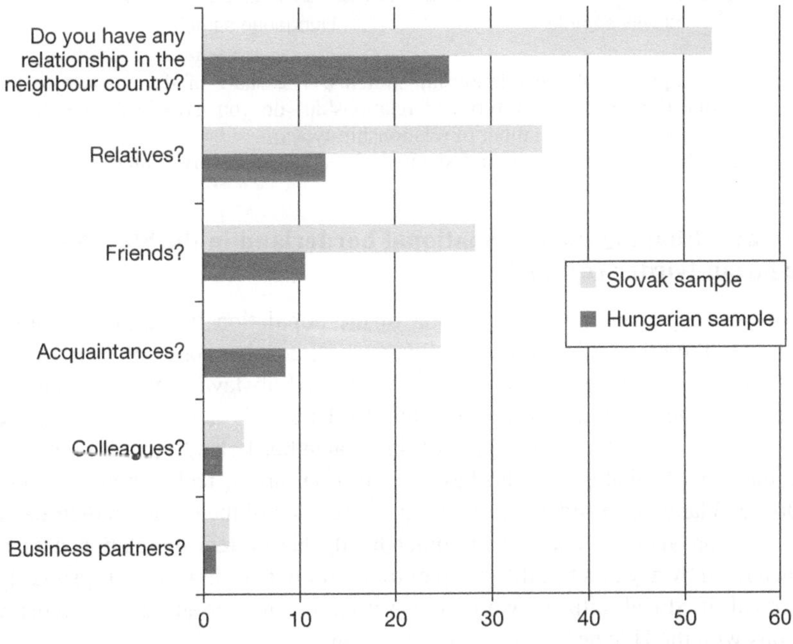

Figure 11.3 Number and type of personal contacts on the other side of the border (share of Yes answers, %).

Source: Figure by the author, based on Csizmadia 2008, questionnaire 2008 (Data).

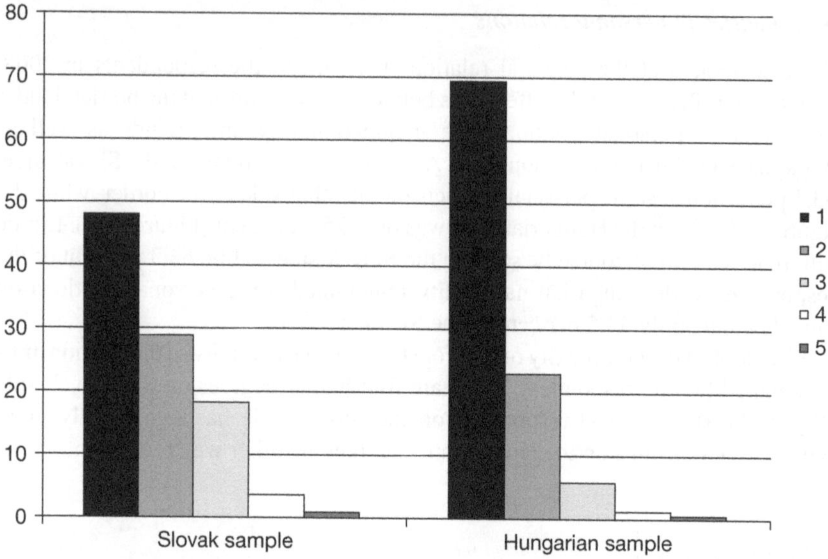

Figure 11.4 Complexity of the relationship systems. The share of the answers to the question 'How many types of relationships do you have in the neighbour country?' (1–5 = number of relationship-types).

Source: Figure by the author, based on Csizmadia 2008, questionnaire 2008 (Data).

Summary: Shaping the transnational borderland in the Slovak–Hungarian border region

As we can see, a significant proportion of the population living in the border region uses the other side with some frequency. Intensive zones of interactions have emerged around some big city junctions (Bratislava, Győr, Komárom/Komarno, Esztergom/Sturovo), where the borderland is now approaching the characteristics of Martinez's category of 'integrated borderland', with a relatively large number of inhabitants who have ties to both areas, feel at home in both cultures and languages and naturally use the other side of the border in their daily lives. We can also see that, on the other hand, this system of relationships is characterized by asymmetry: the group of transnational borderlanders is primarily comprised of Slovak citizens who cherish a larger number and more intensive relations with the Hungarian side than vice versa.

The development of this asymmetry is significantly influenced by ethnic relations, both directly and indirectly. The presence of a Hungarian ethnic group in considerable numbers and proportion on the Slovak side of the border made daily contacts possible and necessary even in the socialist regime when the separating role of the border was much stronger than nowadays. At that time, in addition to visiting relatives there was the possibility of shopping and commuting to work, and so in the socialist system the Slovak–Hungarian border was permeable to an exceptionally high degree. Hungarians on the Slovak side are of course attached to Hungary for emotional, national identity and language reasons as well, and so their relations to Hungary are naturally strong.

If we look at the number and density of interactions, position two is occupied by Slovak ethnic citizens living on the Slovak side. This can be traced back to the indirect effect of the border-crossing ethnic relations. The Slovak population in the Hungarian ethnic environment has a substantial amount of information about the Hungarian people and their culture, including language skills, and so the mental threshold that they have to cross when travelling to the other side of the border is relatively low. This explains why it was members of the Slovak population who first used the opportunity to move from Bratislava to villages on the Hungarian side when it was made possible by the real estate market processes and the permeability of the border. In the villages in the Hungarian border area close to Bratislava 20 to 50 per cent of the inhabitants are now Slovak citizens, many of whom do not speak Hungarian (82 per cent of the movers to Hungary, according to our survey of 2010).

So the group of transnational borderlanders has developed from two sources, as it were: on the one hand, continuing ethno-linguistic relations that were, in part, a counter-effect of the restrictions on minority language use and lifestyle; and, on the other hand, the cross-border effects emerging in the region of Bratislava as a consequence of urban and economic development. In this second context, the main motivation for transborder activities is the practicability of everyday life and the opportunities of the suburban real estate market. While the example of borderlanders in Bratislava could not be characterized by shared ethnic relations, the first variant of borderlanders can be interpreted as a phenomenon that is, to a certain degree, specific to a Central European transnational society and based on a shared cross-border cultural nationality.

Notes

1 Scientific Basic Research for the Socio-economic Analysis of the Slovakian-Hungarian Border Region (ID: HUSKUA/05/02/117). The project was implemented in the Hungary-Slovakia-Ukraine Neighbourhood Programme.

2 AGGLONET Hungarian-Slovak suburbia surrounding Bratislava 2009–2010 supported by Hungary-Slovakia Cross-border Co-operation Programme 2007–2013 (HUSK/0801/1.5.1/0007).

3 The research was supported by the OTKA (Hungarian Scientific Research Fund) 'Analysis of the political geographical spatial structures in the Carpathian Basin (regime changes, cooperation possibilities and absurdities on the turn of the 20th and 21st centuries)', Research leader Tamás Hardi (ID: K 104801).

4 The 'First Vienna Award' (the Vienna Arbitration and Award was a consequence of the Munich Agreement in 1938) separated the ethnic Hungarian-populated territories in southern Slovakia and southern Carpatho-Ukraine from Czechoslovakia and awarded them to Hungary. Slovakia seceded from Czecho-Slovakia in March 1939, and in mid-March 1939 Hungary occupied the rest of Carpatho-Ukraine from the newly founded Slovakia.

Bibliography

Csizmadia, Z., 2008. Társadalmi kapcsolatok a szlovák-magyar határtérségben (Social Contacts in the Slovak–Hungarian Borderland). *Tér és Társadalom*, 22(3), pp.27–51.

Hardi, T., 2001. A határon átnyúló regionális kapcsolatok és együttműködések eltérő lehetőségei. (Different possibilities of cross-border relations and regional cooperation). In Z. Hajdú, N. Pap and J. Tóth, eds., *Az átalakuló Balkán politikai földrajzi kérdései*. Pécs: PTE TTK Földrajzi Intézet Kelet-Mediterrán és Balkán Tanulmányok Központja, pp.88–101.

Hardi, T., ed., 2015. *Terek és tér-képzetek. Elképzelt és formalizált terek a Kárpát-medencében és Közép-Európában.* (Spaces and Space-perceptions. Imaginary and Formalised Spaces in the Carpathian Basin and Central Europe). Somorja: Fórum Kisebbségkutató Intézet.

Hardi, T., Lados, M. and Tóth, K., eds., 2010. *Magyar-szlovák agglomeráció Pozsony környékén – Slovensko-maďarská aglomerácia v okolí Bratislavy.* Győr, Somorja: MTA RKK Nyugat-magyarországi Tudományos Intézet – Fórum Kisebbségkutató Intézet.

Hardi, T., and Tóth, K., eds., 2009. *Határaink mentén: A szlovák-magyar határtérség társadalmi-gazdasági vizsgálata.* Somorja: Fórum Kisebbségkutató Intézet.

Hevesi, A., and Kocsis, K., 2003. *A magyar-szlovák határvidék földrajza.* Dunaszerdahely: Lilium Aurum.

Krejčí, O., 2005. *Geopolitics of the Central European Region. The View from Prague and Bratislava.* Bratislava: VEDA Publishing House of the Slovak Academy of Sciences.

Magocsi, P. R., 2002. *Historical Atlas of Central Europe*. Seattle: University of Washington Press.

Martinez, O. J., 1994. The dynamics of border interaction: New approaches to border analysis. In C. H. Schofield, ed., 1994. *World Boundaries Vol. 1: Global Boundaries.* London and New York: Routledge, pp.1–15.

Index